1,000,000 Books

are available to read at

Forgotten Books

www.ForgottenBooks.com

Read online
Download PDF
Purchase in print

ISBN 978-0-331-73465-2
PIBN 11059183

This book is a reproduction of an important historical work. Forgotten Books uses state-of-the-art technology to digitally reconstruct the work, preserving the original format whilst repairing imperfections present in the aged copy. In rare cases, an imperfection in the original, such as a blemish or missing page, may be replicated in our edition. We do, however, repair the vast majority of imperfections successfully; any imperfections that remain are intentionally left to preserve the state of such historical works.

Forgotten Books is a registered trademark of FB &c Ltd.
Copyright © 2018 FB &c Ltd.
FB &c Ltd, Dalton House, 60 Windsor Avenue, London, SW19 2RR.
Company number 08720141. Registered in England and Wales.

For support please visit www.forgottenbooks.com

1 MONTH OF FREE READING

at

www.ForgottenBooks.com

By purchasing this book you are eligible for one month membership to ForgottenBooks.com, giving you unlimited access to our entire collection of over 1,000,000 titles via our web site and mobile apps.

To claim your free month visit:
www.forgottenbooks.com/free1059183

* Offer is valid for 45 days from date of purchase. Terms and conditions apply.

English
Français
Deutsche
Italiano
Español
Português

www.forgottenbooks.com

Mythology Photography **Fiction**
Fishing Christianity **Art** Cooking
Essays Buddhism Freemasonry
Medicine **Biology** Music **Ancient Egypt** Evolution Carpentry Physics
Dance Geology **Mathematics** Fitness
Shakespeare **Folklore** Yoga Marketing
Confidence Immortality Biographies
Poetry **Psychology** Witchcraft
Electronics Chemistry History **Law**
Accounting **Philosophy** Anthropology
Alchemy Drama Quantum Mechanics
Atheism Sexual Health **Ancient History**
Entrepreneurship Languages Sport
Paleontology Needlework Islam
Metaphysics Investment Archaeology
Parenting Statistics Criminology
Motivational

A NEW SYSTEM
OF
MERCANTILE ARITHMETIC;

ADAPTED TO THE

Commerce of the United States,

IN ITS

DOMESTIC AND FOREIGN RELATIONS;

WITH

FORMS OF ACCOUNTS, AND OTHER WRITINGS USUALLY OCCURRING IN TRADE.

BY MICHAEL WALSH, A. M.

Iter est breve per exempla......SENECA.

THIRD EDITION.

NEWBURYPORT——PRINTED BY E. M. BLUNT, (Proprietor.)
For THOMAS CLARK, Portland.

1804.

Educ T 118.04.874B

John Derby — AD 1805.

HARVARD COLLEGE LIBRARY
BY
GEORGE ARTHUR PLIMPTON
JANUARY 25, 1924

District of Massachusetts District:

··TO WIT:··

BE IT REMEMBERED, That on the seventeenth day of April, in the twenty-fourth year of the Independence of the United States of America, MICHAEL WALSH, of the said District, hath deposited in this Office, the title of a Book, the right whereof he claims as Author, in the words following, to wit:

' A NEW SYSTEM OF MERCANTILE ARITHMETIC: ADAPTED TO THE COMMERCE OF THE UNITED STATES, IN ITS DOMESTIC AND FOREIGN RELATIONS; WITH FORMS OF ACCOUNTS, AND OTHER WRITINGS USUALLY OCCURRING IN TRADE——BY MICHAEL WALSH.'

In conformity to the Act of the Congress of the United States, intituled "An Act for the encouragement of learning, by securing the copies of Maps, Charts and Books, to the Authors and Proprietors of such copies, during the times therein mentioned."

N. GOODALE, Clerk of the District of Massachusetts District.

A true copy of record,

Attest——N. GOODALE, Clerk.

RECOMMENDATIONS.

Newburyport, May 1, 1800.

WE the subscribers having seen Mr. WALSH's New System of MERCANTILE ARITHMETIC, and being satisfied, that it is better calculated, than any yet published, to fit a youth for the business of the Compting-House, cannot but wish it an extensive circulation. The happy elucidation and extended application of the common rules, together with the many original improvements, while they accomplish the student for commerce, are also extremely well adapted to assist and inform the merchant, the mariner, and the trader in their various occupations.

Dudley A. Tyng,
Ebenezer Stocker,
William Bartlet,
Samuel A. Otis, jun.
Tristram Coffin,

Moses Brown,
William Wyer, jun.
Richard Bartlet, jun.
William W. Prout,
Michael Little.

Boston, May 16th, 1800.

WE the subscribers, having examined Mr. WALSH's New System of MERCANTILE ARITHMETIC, and being persuaded that it is better calculated than any we have met with, to qualify young men for admission into Compting-Houses, we wish that it may have an extensive circulation. The clear exemplification and pertinent application of the common rules, together with the many useful additions and improvements which it contains, will render it extremely useful for the merchant, the mariner, and all the other trading classes of society.

Marston Watson,
John C. Jones,
John Codman,
Stephen Higginson,

John Lowell, jun.
Joseph Russell,
Arnold Welles, jun.
Jonathan Jackson.

A 2

Salem, October 7th, 1800.

WE the subscribers, Merchants of Salem, convinced of the necessity of rendering the forms of business, the value of coins, and the nature of commerce, more familiar to the United States as a commercial people, do approve of the MERCANTILE ARITHMETIC of Mr. WALSH, and recommend it as calculated to subserve in the best manner the instruction of our youth, and the purposes of a well-informed merchant.

Wm. Gray, jun.	*Jacob Ashton,*
Benj. Hodges,	*Wm. Prescott,*
B. Pickman,	*Jacob Crowninshield,*
Nath. Bowditch,	*Elias Hasket Derby.*

Preface to the third Edition.

THE merit of WALSH's MERCANTILE ARITHMETIC having been submitted to the public, and established by the most liberal and unequivocal encouragement, the Editor feels a confidence in offering a third Edition of ten thousand copies.

It is unnecessary now to urge the superiority of this over every similar production extant. The discernment of men of letters, and the generous spirit of a commercial public have rendered panegyric useless by an unprecedented patronage. In the very short period of its existence two extensive impressions have been circulated through the country, and orders are already received for a very large proportion of the third.

The value of any work must be decided by those to whom it is more immediately useful; and if such persons possess the means of discrimination the decision will undoubtedly be correct. The present publication is adapted as well to assist the researches of Mathematicians, as to facilitate the negociations of Merchants. Such characters have supported it by their written approbation, and recommended it by an introduction into their own Studies and Compting rooms. Schools and

PREFACE.

Academies have made it the basis of a mercantile education, and it has become an indispensible assistant to every trading class of the community.

This impression has received several valuable additions under the general head of Exchange, including the existing exchange with Antwerp, Trieste, Genoa, Venice, Barcelona, and Palermo in Sicily, and many useful rules under each of these particular heads. A new subject is likewise added, entitled "ARBITRATION OF EXCHANGE," the importance of which will easily be seen by Merchants whose remittances may travel through several countries, and be liable to the rates of Exchange in each.

The errors of the last edition were few and unimportant: But to render the work perfect, they have been minutely considered and corrected.

The Editor is confident that the present edition will be taken up with the same avidity as the two former, and he assures the public that the work shall not suffer, either in accuracy or beauty, by the liberality of its patrons.

EDMUND M. BLUNT.

SEPTEMBER, 1804.

CONTENTS.

	Page
NUMERATION	13
Simple Addition	14
Subtraction	15
Multiplication	15
Division	16
Miscellaneous Questions	19
Table of Money, Weights, Measures, &c.	19
Compound Addition	23
Subtraction	26
Practical Questions in Compound Addition and Subtraction	28
Reduction	29
To find the contents of Grindstones*	32
Reduction of American Monies	34
Compound Multiplication	42
Bills of Parcels	48
Compound Division	49
Decimal Fractions	52
Tables of Coins, Weights and Measures	61
The Single Rule of Three Direct	64
Inverse Proportion	72
Compound Proportion	73
Vulgar Fractions	76
Practice	88
Tare and Tret	95
Single Fellowship	99
Double Fellowship	100
Simple Interest	101
Rule established by the Courts of Law in Massachusetts for making up judgments on securities for Money, which are upon interest, and on which partial payments have been endorsed	116
A Table shewing the number of days, from any day in any month to the same day in any other month through the year	117
Compound Interest	118

* *To find the value see page* 69.

CONTENTS.

	Page
A Table shewing the amount of one pound or one dollar for any number of years under 33, at the rates of 5 and 6 per cent. per annum, compound interest	119
Commission and Brokerage	121
Insurance	123
General Average	124
Buying and Selling Stocks	126
Discount	127
Bank Discount	129
Equation of payments	132
Barter	133
Loss and Gain	135
Alligation Medial	138
Alternate	139
Single Position	142
Double Position	143
Exchange with Great-Britain	146
Ireland	150
Hamburgh	153
Holland	159
Denmark	163
Bremen	165
Antwerp	166
Russia	168
France	170
Tables for changing Livres, Sols and Deniers, to Francs and Centimes	176
Table for reducing Francs and Centimes to Livres, Sols and Deniers	177
Exchange with Spain	178
Barcelona	186
Portugal	188
Leghorn	190
Naples	193
Trieste	194
Genoa	196
Venice	197
Smyrna	198
*Palermo (in Sicily)	273
Jamaica and Bermudas	202
Barbadoes	203
Martinico, Tobago and St. Christopher's	204
French West-Indies	204
Spanish West-Indies	208

* This article ought to follow page 202, but was not received in season. It was handed by a gentleman whose information may be depended on, who arrived a few days previous to the publication of the present edition.

CONTENTS.

	Page
Exchange with Calcutta	210
Bombay	211
Madras	211
Batavia	212
China	214
Manilla	216
Ceylon	216
Japan	217
Tonnage of Goods from the East-Indies to Europe	218
Arbitration of Exchange	220
Mode of calculating American Duties	222
Rates at which all foreign coins are estimated at the Custom-Houses of the United States	225
Arithmetical Progression	226
Geometrical Progression	229
Permutation	232
Extraction of the Square Root	233
of the Cube Root	238
of the Biquadrate Root	243
General Rule for extracting the Roots of all Powers	243
Duodecimals	245
To find the contents of Bales, Cases, &c. in order to ascertain the freight	247
To find ships' tonnage by Carpenter's measure	248
the Government tonnage of ships	251
Tables of Cordage	253
for receiving and paying Gold Coins of France and Spain	255
for receiving and paying Gold Coins of G. Britain and Portugal	256
Mercantile Precedents	257
Bill of Exchange	257
Bill of Goods at an advance on the sterling cost	257
Promissory Note	258
Receipt for an endorsement on a Note	258
for money received on account	258
Promissory Note by two persons	258
General Receipt	258
Bill of Parcels	259
Invoices	260
Accounts of Sales	262
Accounts Current	265
Bill of Sale	269
Interest Account	270
Charter Party	272
Bill of Lading	272
Exchange with Palermo in Sicily	273

EXPLANATION

OF THE CHARACTERS USED IN THIS WORK.

$=$ SIGNIFIES equality, or equal to: as, 20 shillings$=$one pound: that is, 20 shillings are equal to 1 pound.

$+$ Signifies more, or Addition: as $6+6=12$, that is 6 added to 6 is equal to 12.

$-$ Signifies less, or Subtraction; as, $6-2=4$, that is, 6 less 2 is equal to 4.

$\times$ Signifies Multiplication; as, $6\times 2=12$; that is, 6 multiplied by 2 is equal to 12.

$\div$ Signifies Division; as $6\div 2=3$; that is 6 divided by 2 is equal to 3.

Division is sometimes expressed by placing the numbers like a fraction, the upper figure being the dividend, and the lower the divisor: thus, $\frac{54}{6}=9$; that is, 54 divided by 6 is equal to 9.

$: :: :$ Proportion; as, $3 : 6 :: 9 : 18$; that is, as three is to 6, so is 9 to 18.

$\sqrt{}$ Prefixed to any number signifies that the square root of that number is required.

MERCANTILE ARITHMETIC.

ARITHMETIC is the art of computing by numbers, and has five principal rules for this purpose, viz. *Numeration, Addition, Subtraction, Multiplication, and Division.*

NUMERATION

Teacheth to express any proposed number by these ten characters, 0. 1. 2. 3. 4. 5. 6. 7. 8. 9.—0 is called a cypher, and the rest figures or digits. The relative value of which depends upon the place they stand in, when joined together, beginning at the right hand as in the following

TABLE.

9 hundreds of millions.	8 tens of millions.	7 millions.	6 hundreds of thousands.	5 tens of thousands.	4 thousands.	3 hundreds.	2 tens.	1 units.
9	8	7	6	5	4	3	2	1

Though the table consists of only nine places, yet it may be extended to more places at pleasure; as, after hundreds of millions, read thousands of millions, ten thousands of millions, hundred thousands of millions, then millions of millions, &c.

TO WRITE NUMBERS.

RULE. Write down the figures as their values are expressed, and supply any deficiency in the order with cyphers.

SIMPLE ADDITION.

EXAMPLES.

Write down in proper figures the following numbers.
Twenty-nine,
Two hundred and forty-seven,
Seven thousand nine hundred and one,
Eighty-four thousand three hundred and twenty-nine,
Nine hundred and two thousand six hundred and fifteen,
Eighty-nine millions and ninety,
Four millions four hundred thousand and forty,
Nine hundred and nine millions nine hundred and ninety,
Seventy millions seventy thousand and seventy.

Eleven thousand eleven hundred and eleven.	Fourteen thousand fourteen hundred and fourteen.
eleven thousand ・・11000	fourteen thousand・・14000
eleven hundred・・・・1100	fourteen hundred ・・ 1400
eleven ・・・・・・・・・・・・ 11	fourteen ・・・・・・・・・・・・ 14
Total・・12111	Total・・15414

To express in words any number proposed in figures.
RULE. To the simple value of each figure, join the name of its place, beginning at the left hand and reading towards the right.

EXAMPLES.

Write down in words the following numbers.
   46,   199,   2267,   86693,   289732,   1191191,
1169990,   9919,   4320,   55000510.

・・・・・・

SIMPLE ADDITION

Teacheth to collect numbers of the same denomination into one sum.

EXAMPLES.

Gallons.	Yards.	Bushels.
68965	59473	875496
14753	8914	170900
29684	675	574
57693	29	9
171095		
171095		

SIMPLE SUBTRACTION.

Gallons.	Yards.	Bushels.
17573	180041	750010
468	4095	31994
57	83	573
9	7326	74857

As the mercantile method of proving addition is to reckon downwards, as well as upwards, the sums of which will be equal, when the addition is just, two spaces are left for the work.

SIMPLE SUBTRACTION

Teacheth to take a less number from a greater of the same denomination, and thereby to shew the difference.

EXAMPLES.

	Yards.		Gallons.
From	57468532	From	29689141
Take	26587491	Take	17938762
Rem.	30881041	Rem.	11750379
Proof	57468532	Proof	29689141

3 from	924357	take 565383	Rem. 358974
4	517684	291872	225812
5	510090	191939	318151
6	191191	2957	188234
7	291619	829	290790
8	500910	15723	485187

SIMPLE MULTIPLICATION

Is a compendious way of adding numbers of the same name. The number to be multiplied is called the multiplicand. The number which multiplies is called the multiplier. The number arising from the operation is called the product

SIMPLE MULTIPLICATION.

MULTIPLICATION TABLE.

1	2	3	4	5	6	7	8	9	10	11	12
2	4	6	8	10	12	14	16	18	20	22	24
3	6	9	12	15	18	21	24	27	30	33	36
4	8	12	16	20	24	28	32	36	40	44	48
5	10	15	20	25	30	35	40	45	50	55	60
6	12	18	24	30	36	42	48	54	60	66	72
7	14	21	28	35	42	49	56	63	70	77	84
8	16	24	32	40	48	56	64	72	80	88	96
9	18	27	36	45	54	63	72	81	90	99	108
10	20	30	40	50	60	70	80	90	100	110	120
11	22	33	44	55	66	77	88	99	110	121	132
12	24	36	48	60	72	84	96	108	120	132	144

EXAMPLES.

Multiplicand	5965468	4765293	6281947
Multiplier	2	3	4
Product	11930936	14295879	25127788

		by		product	
4 Mult.	2658758		5		13293790
5	9674372		6		58046232
6	7689657		7		53827599
7	2674876		9		24073884
8	4198543		10		41985430
9	7491685		11		82408535
10	2689489		12		32273868
11	1768735		20		35374700
12	2891496		400		1156598400
13	5749857		78		448488846
14	2653294		872		2313672368
15	78965987		5893		465346561391
16	562910859		490070		275868665090130

. . . . . .

SIMPLE DIVISION

Teacheth to find how often one number is contained in another of the same name.

The number given to be divided, is called the *dividend.*

The number by which to divide, is called the *divisor.*

SIMPLE DIVISION.

The number of times the *divisor* is contained in the *dividend*, is called the *quotient*.

The *remainder*, if there be any, will be less than the *divisor*.

PROOF.

Multiply the quotient by the divisor; to the product add the remainder, and the sum will be equal to the dividend, if the work be right.

EXAMPLES.

```
           Dividend
Divisor   2)694568946·            3)2768954684
           ──────────             ────────────Rem.
Quotient   347284473               922984861—1
                   2                       3
           ──────────             ────────────
Proof      694568946               2768954584
```

```
           Dividend.. Quotient..
Divisor.  52)6495436(124912
             52           52
             ──          ─────
            129          249824
            104          .624560
            ───                 12 Rem.
            255          ────────
            208          6495436 Proof..
            ───
            474
            468
            ───
             63
             52
            ───
            116
            104
            ───
             12
```

SIMPLE DIVISION.

	Divide	by		Quotient.	Rem.
4	8965462	6	Ans.	1494243 and	4
5	3728675	8		466084	3
6	4654682	9		517186	8
7	2768967	10		276896	7
8	1949952	11		177268	4
9	2968967	12		247413	11
10	5268794	20		263439	14
11	29619145	40		740478	25
12	419825367	500		839650	367
13	296876234	64		4638691	10
14	47989536925	735		65291886	715
15	26574983184	8962		2965296	432
16	53479689236	7684		6959876	2052
17	4917968967	2359		2084768	1255
18	3258675689	67435		48323	14184

When the divisor is a compound number, that is, if any two figures, being multiplied together, will make that number, then divide the dividend by one of those figures, and the first quotient by the other figure, and it will then give the quotient required.—But as it sometimes happens that there is a remainder to each of the quotients, and neither of them the true one, it may be found by this

RULE. Multiply the first divisor by the last remainder, and to the product add the first remainder, which will give the true one.

EXAMPLES.

Divide 296876234, by 64
8)296876234
————————
8)37109529—2
————————
Quotient 4638691 and 1 × 8 + 2 = 10 remaining.

Divide 8757635 by 28   Divide 18957492 by 42
————————   ————————
Quotient 312772 and 19 rem.   451368 and 36 rem.

Divide 1571196 by 72   Divide 3751749 by 96
————————   ————————
Quotient 21822 and 12 rem.   39080 and 69 rem.

MONEY, WEIGHTS, MEASURES, &c.

MISCELLANEOUS QUESTIONS.

1. Add 562163, 21964, 56321, 18536, 4340, 279, and 83 together.   Ans. 663686.

2. What number is it, which being added to 9709 will make 110901?   Ans. 101192.

3. General WASHINGTON was born in the year 1732; how old was he in 1799?   Ans. 67 years.

4. Add up twice 397, three times 794, four times 3176, five times 15880, six times 95280, and once 333040.
  Ans. One Million.

5. A cashier received, viz. Four hundred and nine dollars, Twenty thousand and thirteen dollars, Eight thousand five hundred and ten dollars, Nine hundred and twenty-eight dollars; of which he paid away Fifteen thousand fifteen hundred and fifteen dollars: What was the whole sum he received, and how much remains after deducting the payment?
Ans. He received 29860 dolls. and there remains 13345 dolls.

6. What is the product of 15927 multiplied by 4009?   Ans. 63851343.

7. 128 men have one half of a prize, worth 34560 dollars, to be equally divided between them: What is each man's part?   Ans. 135 dollars.
  * Prove this answer to be right.

8. Three merchants, A, B, and C, have a stock of 14876 dollars, of which A put in 4963 dolls. B 5188 dolls. and C the remainder: How much did C put in?   Ans. 4715 dollars.

....

TABLE OF MONEY, WEIGHTS, MEASURES, &c.

FEDERAL MONEY.

10 Mills .... make .... 1 Cent.
10 Cents .... 1 Dime.
10 Dimes, or 100 Cents .... 1 Dollar.
10 Dollars .... 1 Eagle.

ENGLISH MONEY.

4 Farthings .... make .... 1 Penny.
12 Pence .... 1 Shilling.
20 Shillings .... 1 Pound.

MONEY, WEIGHTS, MEASURES, &c.

PENCE TABLE.			SHILLINGS TABLE.		
d.	s.	d.	s.	£.	s.
20 are	1	8	20 are	1	0
30	2	6	30	1	10
40	3	4	40	2	0
50	4	2	50	2	10
60	5	0	60	3	0
70	5	10	70	3	10
80	6	8	80	4	0
90	7	6	90	4	10
100	8	4	100	5	0
110	9	2	110	5	10
120	10	0	120	6	0
130	10	10	130	6	10
140	11	8	140	7	0
150	12	6	150	7	10
200	16	8	200	10	0

TROY WEIGHT.

24 Grains .... make .... 1 Pennyweight.
20 Pennyweights .... 1 Ounce.
12 Ounces .... 1 Pound.

NOTE. By this weight are weighed jewels, gold, silver and liquors.

AVOIRDUPOIS WEIGHT.

16 Drams .... make .... 1 Ounce.
16 Ounces .... 1 Pound.
28 Pounds .... 1 Quarter.
4 Quarters .... 1 Hundred weight.
20 Hundred weight .... 1 Ton.

NOTE. By this weight are weighed such commodities as are coarse and subject to waste, and all metals, except gold and silver. One pound Avoirdupois is equal to 14 oz. 11 pwt. and 15$\frac{1}{2}$ grs. Troy.

APOTHECARIES WEIGHT.

20 Grains .... make .... 1 Scruple.
3 Scruples .... 1 Dram.
8 Drams .... 1 Ounce.
12 Ounces .... 1 Pound.

NOTE. Apothecaries use this weight in compounding their medicines; but they buy and sell their drugs by Avoirdupois weight.

CLOTH MEASURE.

4 Nails .... make .... 1 Quarter.
4 Quarters .... 1 Yard.
3 Quarters .... 1 Ell Flemish.
5 Quarters .... 1 Ell English.
6 Quarters .... 1 Ell French.

MONEY, WEIGHTS, MEASURES, &c.

Long Measure.

3 Barley Corns ······ make ······ 1 Inch.
12 Inches ·········· 1 Foot.
3 Feet ·········· 1 Yard.
5½ Yards, or 16½ Feet ·········· 1 Pole, Rod, or Perch.
40 Poles ·········· 1 Furlong.
8 Furlongs ·········· 1 Mile.
3 Miles ·········· 1 League.
60 Geographical, or }
69½ Statute Miles } ·········· 1 Degree.

NOTE. In this measure, length only is considered.

Land or Square Measure.

144 Square Inches···· make···· 1 Square Foot.
9 Feet ·········· 1 Yard.
30¼ Yards, or }
272¼ Feet } ·········· 1 Pole, Rod, or Perch.
40 Poles or Perches ·········· 1 Rood.
4 Roods ·········· 1 Acre.

NOTE. This measure respects length and breadth.

Wine Measure.

2 Pints ········ make ········ 1 Quart.
4 Quarts ·········· 1 Gallon.
42 Gallons ·········· 1 Tierce.
63 Gallons ·········· 1 Hogshead.
84 Gallons ·········· 1 Puncheon.
2 Hogsheads ·········· 1 Pipe or Butt.
2 Pipes or 4 Hogsheads ······ 1 Tun.

NOTE. The wine gallon contains 231 cubic inches.

Ale and Beer Measure.

2 Pints ········ make ·········· 1 Quart.
4 Quarts ·········· 1 Gallon.
8 Gallons ·········· 1 Firkin of Ale.
9 Gallons ·········· 1 Firkin of Beer.
2 Firkins ·········· 1 Kilderkin.
2 Kilderkins ·········· 1 Barrel.
54 Gallons ·········· 1 Hhd. of Beer.
3 Barrels ·········· 1 Butt.

NOTE. The ale gallon contains 282 cubic inches.

MONEY, WEIGHTS, MEASURES, &c.

CUBIC OR SOLID MEASURE.

1728 Inches .... make .... 1 Foot.
27 Feet .... 1 Yard.
40 Feet of round Timber or ⎫
50 Feet of hewn Timber   ⎭ ...... 1 Ton or Load.
128 Solid Feet .... 1 Cord of Wood.

NOTE. 8 feet in length, 4 in breadth, and 4 in height, making 128 solid feet, contain a cord of wood. This measure respects length, breadth and thickness.

DRY MEASURE.

2 Pints .... make .... 1 Quart.
2 Quarts .... 1 Pottle.
2 Pottles .... 1 Gallon.
2 Gallons .... 1 Peck.
4 Pecks .... 1 Bushel.
2 Bushels .... 1 Strike.
4 Bushels .... 1 Coom.
8 Bushels .... 1 Quarter.
36 Bushels .... 1 Chaldron.
5 Quarters .... 1 Wey.
2 Weys .... 1 Last.

NOTE. The gallon dry measure contains $268\frac{4}{5}$ cubic inches.

TIME.

60 Seconds .... make .... 1 Minute.
60 Minutes .... 1 Hour.
24 Hours .... 1 Day.
365 Days .... 1 Year.

NOTE. 365 days 5 hours 48 minutes 57 seconds make a solar year, according to the most exact observation.

The number of days in each month is thus found:

*Thirty days hath September, April, June, and November;
February hath twenty-eight alone, and all the rest have thirty-one.*

When the year can be divided by 4 without a remainder, it is Bissextile or Leap Year, in which February hath 29 days.

COMPOUND ADDITION

Teacheth to collect numbers of different denominations into one total.

FEDERAL MONEY.

D.	C.	M.	D.	C.	M.
174	74	3	396	14	4
198	19	3	147	19	5
157	14	4	149	57	9
196	76	9	157	83	8

ENGLISH MONEY.

£.	s.	d.	£.	s.	d.
149	14	$6\frac{3}{4}$	814	16	$6\frac{1}{4}$
387	19	$8\frac{1}{2}$	198	18	$8\frac{3}{4}$
259	16	$7\frac{1}{2}$	376	14	$9\frac{1}{2}$
874	17	$4\frac{3}{4}$	226	16	$7\frac{3}{4}$
678	15	$6\frac{1}{2}$	174	17	$10\frac{1}{2}$

TROY WEIGHT.

lb.	oz.	dwt.	gr.	lb.	oz.	dwt.	gr.
48	7	14	19	83	11	15	22
95	4	17	22	15	6	16	19
27	5	14	15	21	8	19	23
65	6	19	16	33	9	15	14
19	7	13	15	46	4	13	17

AVOIRDUPOIS WEIGHT.

Ton.	Cwt.	qr.	lb.	oz.	dr.	Cwt.	qr.	lb.
18	17	1	14	13	13	593	1	19
36	15	3	16	13	15	187	3	19
29	15	2	19	12	13	159	2	25
14	16	3	27	14	12	283	3	13
16	19	2	25	13	10	146	2	18
57	17	1	14	15	9	259	1	22

COMPOUND ADDITION.

APOTHECARIES' WEIGHT.

lb.	oz.	dr.	sc.	gr.		lb.	oz.	dr.	sc.	gr.
3	7	5	1	17		2	5	3	2	11
1	3	2	2	13		1	2	2	1	14
2	5	3	2	14		3	3	5	2	13
3	4	2	1	15		5	5	4	1	12
5	2	2	2	17		2	9	3	2	15
2	3	1	2	18		1	6	4	2	17

CLOTH MEASURE.

yd.	qr.	nl.	E.Fl.	qr.	nl.	E.Fr.	qr.	nl.	E.E.	qr.	nl.
571	1	3	873	2	3	181	2	2	56	1	2
184	2	2	196	2	2	196	3	3	19	2	3
196	2	3	158	1	1	157	4	2	14	3	2
283	3	2	147	2	3	168	3	3	26	4	3
146	2	3	326	2	2	193	5	2	83	2	2
375	3	2	194	2	1	214	2	3	57	3	3

WINE MEASURE.

Tun.	hhd.	gal.	qt.	pt.		Tun.	hhd.	gal.	qt.	pt.
187	1	17	3	1		176	3	16	2	1
56	3	15	2	1		59	2	57	3	1
9	1	29	3	1		8	3	14	2	1
36	2	18	2	1		17	2	19	1	1
217	3	57	1	1		168	1	38	2	1
56	1	46	2	1		25	2	52	3	1

ALE AND BEER MEASURE.

hhd.	gal.	qt.	pt.		hhd.	gal.	qt.	pt.
49	38	2	1		78	17	3	1
38	45	3	1		19	16	2	1
57	48	2	1		15	51	3	1
49	37	1	1		76	43	2	1
57	26	2	1		23	26	3	1
28	18	3	1		52	38	2	1

COMPOUND ADDITION.

DRY MEASURE.

qr.	bush.	pck.	qt.
57	4	2	1
19	5	3	1
38	6	2	3
27	7	3	7
5	3	1	4
9	2	2	3
72	5	3	2

chal.	bush.	pck.	qt.
576	31	1	3
19	27	2	2
56	15	3	5
25	8	2	4
9	9	1	6
14	15	2	3
32	26	3	2

LONG MEASURE.

deg.	mil.	fur.	po.	ft.	in.	bar.
217	17	7	19	14	9	1
733	17	4	16	13	3	2
283	53	5	19	12	2	2
346	26	6	23	13	4	1
189	32	3	27	14	5	2
176	14	2	15	15	6	2
921	15	4	18	16	7	1

mil.	fur.	po.	yd.	ft.
876	7	13	4	2
129	6	26	2	1
167	4	19	3	2
157	3	15	2	2
286	2	27	1	1
194	5	32	2	2
176	4	18	5	2

LAND MEASURE.

acr.	roo.	per.
741	1	19
69	3	29
15	2	16
37	3	14
16	2	13
29	3	27

acr.	roo.	per.
870	3	19
19	2	16
54	3	37
129	2	26
187	3	14
136	2	19

TIME.

yrs.	days.	hrs.	min.	sec.
187	149	14	13	12
146	126	16	16	16
59	186	19	39	19
28	140	21	46	35
7	119	22	18	26
146	146	19	57	19

yrs.	days.	hrs.	min.	sec.
300	169	14	16	17
19	186	17	16	16
46	147	15	19	19
87	196	23	46	47
157	219	14	23	16
46	138	15	42	13

C

COMPOUND SUBTRACTION

Teacheth to find the inequality between numbers of divers denominations.

FEDERAL MONEY.

	dol.	ct.	m.		dol.	ct.	m.		dol.	ct.	m.
From	1901	95	1		435	00	1		170	10	3
Take	992	97	2		9	15	9		9	50	2

ENGLISH MONEY.

	£.	s.	d.		£.	s.	d.
From	191	11	3½		304	19	8¼
Take	114	16	2¼		126	16	8½

	£.	s.	d.		£.	s.	d.
From	389	18	0½		100	0	5
Take	9	19	4		11	11	2¾

TROY WEIGHT.

	lb.	oz.	dwt.	gr.		lb.	oz.	dwt.	gr.
From	87	11	11	13		27	10	15	22
Take	19	11	14	22		15	9	16	23

AVOIRDUPOIS WEIGHT.

	ton.	cwt.	qr.	lb.	oz.	dr.		cwt.	qr.	lb.
From	100	10	1	11	14	13		59	1	11
Take	15	13	1	18	12	15		19	3	27

APOTHECARIES' WEIGHT.

	lb.	oz.	dr.	sc.	gr.		lb.	oz.	dr.	sc.	gr.
From	2	3	4	1	13		2	1	3	1	15
Take	1	7	5	2	10		1	4	2	2	17

COMPOUND SUBTRACTION.

CLOTH MEASURE.

	yd.	qr.	nl.	E.Fl.	qr.	nl.	E.E.	qr.	nl.	E.Fr.	qr.	nl.
From	251	1	2	189	2	1	419	1	3	389	2	2
Take	127	3	3	120	2	2	174	3	2	189	5	3

WINE MEASURE.

	tun.	hhd.	gal.	qt.	pt.	tun.	hd.	gal.	qt.	pt.
From	591	1	13	1	1	800	1	50	2	1
Take	126	2	56	3	1	149	2	61	3	1

ALE and BEER MEASURE.

	hd.	gal.	qt.	pt.	hd.	gal.	qt.	pt.
From	571	19	3	1	100	36	2	1
Take	198	53	2	1	9	27	3	1

DRY MEASURE.

	qr.	bu.	gal.	qt.	chal.	bu.	gal.	qt.
From	38	4	5	3	69	21	3	2
Take	17	5	1	2	49	33	5	3

LONG MEASURE.

	deg.	m.	fur.	p.	f.	in.	b.	m.	fur.	p.	f.
From	819	13	1	19	11	3	1	219	3	14	11
Take	159	49	2	27	16	8	2	209	7	15	12

LAND MEASURE.

	acr.	roo.	per.	acr.	roo.	per.	acr.	roo.	per.
From	591	1	11	501	3	13	219	2	21
Take	129	3	15	190	2	21	156	1	36

TIME.

	yrs.	da.	hr.	m.	sec.	yrs.	da.	hr.	m.	sec.
From	171	143	11	14	19	811	111	15	23	52
Take	128	174	19	51	14	389	190	21	48	54

PRACTICAL QUESTIONS IN COMPOUND ADDITION AND SUBTRACTION.

1. Cast up the following sums, viz. twenty-three shillings and five pence, one pound and nine pence, seven shillings and eleven pence three farthings, twenty pounds thirteen shillings and nine pence, fifteen pence three farthings.

£.	s.	d.
1	3	5
1	0	9
0	7	11¾
20	13	9
0	1	3¾

Ans. £. 23 7 2½

Proof £. 23 7 2½

2. Twenty dollars and four cents, five dollars and three mills, eighty-two cents, six dollars and five mills.
Ans. 31 dols. 86 cts. 8 m.

3. Seventy dollars, three dollars and three cents, thirty-four cents and four mills, eighty dollars and a half, six dollars and a quarter. Ans. 160 dols. 12 cts. 4 mills.

4. Ten pounds and three pence, forty-five shillings and ten-pence half penny, thirty-seven shillings and four-pence three farthings, nine pounds and three farthings, one shilling and six pence farthing, eighty-two shillings and four-pence half-penny. Ans. £.27 7 5¾

5. Thirty dollars six cents and a half, fifty-three cents and three quarters, eleven cents and a quarter, nine dollars eleven cents and a half, fifty-four cents. Ans. 40 dols. 37 cents.

6. Take three shillings and four pence from one pound two shillings and a penny. Ans. 18s. 9d.

7. From £.5 2s. 1d. take nine shillings and six-pence half-penny. Ans. £.4 12 6½

8. Take twenty shillings and three farthings from £.8. Ans. £.6 19 11¼

9. From 18 dollars take eight mills. Ans. 17 dols. 99 cts. 2 m.

10. Take 53 dimes from 53 eagles. Ans. 524 dols. 7 dimes or 70 cts.

11. A merchant bought 112 bars of iron, weighing 56 cwt. 1 qr. 11 lb. of which he sold 59 bars, weighing 29 cwt. 3 qrs.

REDUCTION.

21 lb.; how many bars has he remaining, and what is the weight?   Ans. 53 bars, weighing 26 cwt. 1 qr. 18 lb.

12. Required the total weight of 4 hogsheads of sugar, weighing as follows, viz. No. 1. 9 cwt. 2 qrs. 21 lb. No. 2. 10 cwt. 3 qrs. 23 lb. No. 3. 8 cwt. 2 qrs. 25 lb. No. 4. 9 cwt. 3 qrs. 17 lb.   Ans. 39 cwt. 1 qr. 2 lb.

13. A ropemaker received 3 tons 15 cwt. 3 qrs. 14 lb. of hemp to be wrought, of which he delivered in cordage 34 cwt. 1 qr. 22 lb.; how much remains?   Ans. 2 tons 1 cwt. 1 qr. 20 lb.

14. Received 57953 mills, 4953 cents, 1913 dimes, and 45 eagles; required the total sum?   Ans. 748 dols. 78 cts. 3 mills.

15. A cashier received, viz. one hundred pounds and nine-pence half-penny, three thousand seven hundred and four pounds ten shillings, twenty thousand and ninety pounds two shillings and eleven pence three farthings, of which he paid away sixteen thousand sixteen hundred and sixteen pounds; how much has he on hand?   Ans. £.6278 13. 9¼

16. A farmer bought three pieces of land, measuring, viz. the first piece 21 acres 3 roods 19 poles; the second, 37 acres 2 roods 29 poles; the third, 27 acres 2 roods 25 poles; of which he sells 15 acres 2 roods 39 poles; how much has he remaining?   Ans. 71 acres 1 rood 34 poles.

17. A has paid B £.9 15 6¼, £.19 11 9¾, £.14 19 7½, and 54s. 3¼d. on account of a debt of £.50; how much is there still unpaid?   Ans. £.2 18 9¼

REDUCTION.

REDUCTION teacheth to change numbers from one denomination to another, without losing their value.

RULE. When the Reduction is descending, multiply the highest denomination by as many of the next less as make one of the greater, adding to the product the parts of the same name, and so on to the last.

When the Reduction is ascending, divide the given number by as many of that denomination as make one of the next higher, and so on to the denomination required, and the last quotient with the several remainders, (if any) will be the answer.

The *proof is by reversing the question.*

REDUCTION.

FEDERAL MONEY.

1. In 53 dollars how many mills?

    53 dolls.   ⎧ Or decimally, by adding a cypher
       10   ⎬ for each inferior denomination, thus,

   530 dimes. ⎩
     10

  5300 cents.
    10
                             dol.d.c.m.
Ans. 53000 mills.   53,000

2. In 14000 mills how many dollars?

  10)14000   ⎧ Or decimally, by separating the figures,
   —————   counting from the right to the name
   10)1400   ⎨ required, thus,

    10)140   ⎩
     ————   *dol.d.c.m.*
   Ans. 14 dolls.   14,000

3. In 57935 mills how many dollars?
               Ans. 57 dollars, 93 cents, and 5 mills.
4. How many eagles in 1933 dimes?
               Ans. 19 eagles, 3 dollars, 3 dimes.
5. In 1290 mills how many dimes?
               Ans. 12 dimes and 9 cents.
6. How many cents in 46 dollars?   Ans. 4600.
7. In 190004 mills how many dollars?
               Ans. 190 dollars and 4 mills.

ENGLISH MONEY.

1. In £.91   11   3½ how many farthings?
        91   11   3½   Proof.
        20   4)87902

    1831 shillings.   12)21975—2
       12

   21975 pence.   20)1831—3
        4
                              £.91   11   3½

Ans. 87902 farthings.

2. How many pounds in 3175 farthings?   Ans. £.3   6   1¾

REDUCTION.

3. In 19s. 8¾d. how many farthings? Ans. 947 farthings.
4. How many pounds in 9752 pence? Ans. £.40 12 8
5. In £.46 how many crowns of 6s. 7d. each?
   Ans. 139 crowns and 4 shillings and 11 pence.
6. How many pounds in 493 dollars? Ans. £.147 18
7. In 143 pence, how many shillings? Ans. 11s. 11d.
8. Reduce 38s. 4½d. to half pence. Ans. 921 half pence.

Prove the above answers to be right.

TROY WEIGHT.

1. In 15lb. troy how many grains? Ans. 86400 grs.
2. How many ounces in 5749 dwt.? Ans. 287 oz. 9 dwt.
3. In 11 oz. 13 dwt. 13 grs. how many grains?
   Ans. 5605 grs.
4. How many grains in 15 spoons, each weighing 6 dwt. 15 grs.? Ans. 2385 grs.

AVOIRDUPOIS WEIGHT.

1. In 19 tons 14 cwt. 2 qrs. 19 lb. 11 oz. 13 drs. how many drams? Ans. 11316157 drs.
2. How many cwt. in 9563 lb.?
   Ans. 85 cwt. 1 qr. 15 lb.
3. In 13 cwt. 3 qrs. 21 lb. how many pounds?
   Ans. 1561 lb.
4. How many mess-pieces of 4½ lb. and 3½ lb. of each an equal number, in 31 cwt. 1 qr. 12 lb. of beef?
   Ans. 439 pieces of each.

WINE MEASURE.

1. In 25 tuns of wine how many pints? Ans. 50400 pints.
2. How many hogsheads in 4935 quarts?
   Ans. 19h. 36g. 3 qt.
3. In 3 hhds. 13 gals. 2 qts. how many half pints?
   Ans. 3240 half pints.

CLOTH MEASURE.

1. In 158 yards how many nails? Ans. 2528 nails.
2. How many ells English in 5932 nails?
   Ans. 296 ells 3 qrs.
3. In 29 pieces of holland, each containing 36 ells Flemish, how many yards? Ans. 783 yds.

REDUCTION.

Long Measure.

1. In 29 miles how many inches? Ans. 1837440 inches.
2. How many furlongs in 19753 yards?
Ans. 89 fur. 173 yds.
3. In 590057 inches how many leagues?
Ans. 3 leag. 2 fur. 110 yds. 1 f. 5 in.

Time.

1. How many hours in 57 years, allowing each year to be 365 days 6 hours? Ans. 499662 hours.
2. In 57953 hours how many weeks?
Ans. 344 w. 6 da. 17 hr.
3. How many days from 19th of March to the 23d September following? Ans. 188 days.
4. How many days from 24th May, 1797, to 15th December, 1798? Ans. 570 days.

Land Measure.

1. In 41 acres 2 roods 14 perches, how many rods?
Ans. 6654 rods or perches.
2. How many square rods in 7752 square feet?
Ans. 28 rods, 129 feet.
3. In 5972 perches, how many acres?
Ans. 37 ac. 1 rood 12 per.

Solid Measure.

1. In a pile of wood 96 feet long, 5 feet high, and 4 feet wide, how many cords? Ans. 15 cords.
2. In 82 tons of round timber how many inches?
Ans. 5667840 inches.
3. What are the contents of a load of wood, 6 feet long, 4 feet high, and 2¼ feet wide? Ans. 3¾ feet.

GRINDSTONES are sold by the cubic foot, commonly called a stone, and the contents are thus found :—

RULE. To the whole diameter add half of the diameter, and multiply the sum of these by the same half, and this product by the thickness; divide this last number by 1728, the inches in a cubic foot, and the quotient is the contents, or answer required.

REDUCTION.

EXAMPLES.

4. How many cubic feet in a grindstone, 24 inches diameter, and 4 inches thick?

```
        24 diameter.
        12 half diameter.
        ――――
        36
        12
        ――――
        432
         4 thickness.
        ――――
   1728)1728
```

Ans. 1 foot.

5. What are the contents of a grindstone, 36 inches diameter, and 4 inches thick?

```
          36
          18
         ――――
          54
          18
         ――――
         432
          54
         ――――
         972
           4
      ――――――――
   1728)3888(2¼
        3456
        ――――
         432
           4
      ――――――――
   1728)1728(1
        1728
        ――――
```

Ans. 2¼ cubic feet.

REDUCTION.

AMERICAN MONIES.

To change New-England and Virginia currency to Federal money, the dollar being 6 shillings.

RULE. As the value of a dollar is equal to three tenths of a pound, when pounds are given to be changed, annex three cyphers to the sum, and divide the whole by 3; the quotient is the answer in cents.

EXAMPLES.

1. Change £.523 to Federal money.

3)523000

17433⅓ cents. Ans. 1743 dols. 33⅓ cts.

Change the following sums, viz.

	£.		dols. cts.
2.	184	Ans.	613 33⅓
3.	29		96 66⅔
4.	57		190
5.	219		730
6.	81		270
7.	127		423 33⅓

When pounds and shillings are given, to the pounds annex half the number of shillings and two cyphers, if the number of shillings in the given sum be even; but if the number be odd, annex half the number, and then 5 and one cypher, and divide by 3; the quotient is the answer in cents.

EXAMPLES.

1. Change £.59 18s. to Federal money.

3)59900

19966⅔ cts. Ans. 199 dols. 66⅔ cts.

2. Change £.93 13s. to Federal money.

3)93650

31216⅔ cts. Ans. 312 dols. 16⅔ cts.

Change the following sums, viz.

	£.	s.		dols. cts.
3.	129	13	Ans.	432 16¾
4.	68	15		212 50
5.	27	18		93
6.	182	19		609 83⅓
7.	57	16		192 66⅔
8.	121	7		404 50

REDUCTION.

When there are shillings, pence, &c. in the given sum, annex for the shillings as before directed, and to these add the farthings in the given pence and farthings, observing to increase their number by one when they exceed 12, and by two when they exceed 37, and divide as before.

EXAMPLES.

1. Change £.21 2s. 4½d. to Federal money.

3)21419

7139⅔ cts.

4 is annexed to the pounds for half the shillings, and 19 for the farthings in 4½d. and excess of 12.

Ans. 71 dols. 39⅔ cts.

2. Change £.117 16s. 2d. to Federal money.

3)117808

39269⅓ cts.　Ans. 392 dols. 69⅓ cts.

3. Change £.721 9s. 11¼d. to Federal money.

3)721497

240499 cts.

In this example 4 is annexed to the pounds for half the even shillings, and 47 for the farthings in 11¼d. and excess of 37, and then 5 is added to the figure next to half the shillings, making it 9 in place of 4 for the odd shilling.

Ans. 2404 dols. 99 cts.

4. Change £.29 11s. 2¼d. to Federal money.

3)29559

9853 cts.　Ans. 98 dols. 53 cts.

Change the following sums, viz.

	£.	s.	d.	dols.	cts.
5.	25	19	9	Ans. 86	62½
6.	24	11	7¾	81	94
7.	1238	10	9½	4128	46⅔
8.	2001	1	3½	6670	21¾
9.	153	17	6	512	91½

REDUCTION.

A TABLE

FOR CHANGING SHILLINGS AND PENCE INTO CENTS AND MILLS.

pence.	shill. 0 cts. m.	shill. 1 cts. m.	shill. 2 cts. m.	shill. 3 cts. m.	shill. 4 cts. m.	shill. 5 cts. m.
0		16 7	33 3	50 0	66 7	83 3
1	1 4	18 1	34 7	51 4	68 1	84 7
2	2 8	19 5	36 1	52 8	69 5	86 1
3	4 2	20 9	37 5	54 2	70 9	87 5
4	5 6	22 3	38 9	55 6	72 3	88 9
5	7 0	23 7	40 3	57 0	73 7	90 3
6	8 3	25 0	41 7	58 3	75 0	91 7
7	9 7	26 4	43 0	59 7	76 4	93 0
8	11 1	27 8	44 4	61 1	77 8	94 4
9	12 5	29 2	45 8	62 5	79 2	95 8
10	13 9	30 6	47 2	63 9	80 6	97 2
11	15 3	32 0	48 6	65 3	82 0	98 6

To change Federal Money to New-England and Virginia Currency.

RULE. When the sum is dollars only, multiply it by 3 and double the first figure of the product for shillings, and the rest of the product will be pounds.

When there are cents in the given sum, multiply the whole by 3, and cut off three figures of the product to the right hand as a remainder.

Multiply this remainder by 20 and cut off as before.

Proceed in this manner through the several parts of a pound, and the numbers standing on the left hand, make the answer, in the several denominations.

NOTE. If there be mills, cut off four figures and proceed as above.

EXAMPLE.

1. Change 872 dollars to New-England currency.

$$872 \times 3 = 261\ 12$$

£. s.
Ans. 261 12

2. Change 1971 dols. 96⅔ cts. to Massachusetts currency.

```
  1971 96⅔
         3
  ---------
£.591,590
       20
  ---------
s. 11,800
       12
  ---------
d. 9,600
        4
  ---------
f. 2,400
```

Ans. £.591 11 9½

3. Reduce 1259 dols. 89 cts. and 7 mills, to Mass. currency.

```
  1259 89 7
          3
  ----------
£.377,9691
        20
  ----------
s. 19,3820
        12
  ----------
d. 4,5840
         4
  ----------
f. 2,3360
```

Ans. £.377 19 4½

A TABLE

For changing Cents into Shillings, Pence, and Farthings.

		Cents. 10		Cents. 20		Cents. 30		Cents. 40		Cents. 50		Cents. 60		Cents. 70		Cents. 80		Cents. 90	
cents	*d.*	*s.*	*d.*	*s.*	*d.*	*s.*	*d.*	*s.*	*d.*	*s.*	*d.*	*s.*	*d.*	*s.*	*d.*	*s.*	*d.*	*s.*	*d.*
0			7¼	1	2½	1	9½	2	4¾	3	0	3	7¼	4	2½	4	9½	5	4¾
1	¾		8	1	3	1	10¼	2	5½	3	0¾	3	8	4	3	4	10¼	5	5½
2	1½		8¾	1	3¾	1	11	2	6¼	3	1½	3	8¾	4	3¾	4	11	5	6¼
3	2¼		9¼	1	4½	1	11¾	2	7	3	2¼	3	9¼	4	4½	4	11¾	5	7
4	2¾		10	1	5¼	2	0½	2	7¾	3	2¾	3	10	4	5¼	5	0½	5	7¾
5	3½		10¾	1	6	2	1¼	2	8½	3	3½	3	10¾	4	6	5	1¼	5	8½
6	4¼		11½	1	6¼	2	2	2	9	3	4¼	3	11½	4	6¾	5	2	5	9
7	5	1	0¼	1	7½	2	2¾	2	9¾	3	5	4	0¼	4	7½	5	2¾	5	9¾
8	5¾	1	1	1	8	2	3¼	2	10½	3	5¾	4	1	4	8	5	3¼	5	10½
9	6½	1	1¾	1	8¾	2	4	2	11¼	3	6½	4	1¾	4	8¾	5	4	5	11¼

To change New-York and North-Carolina currency to Federal money, the dollar being 8 shillings.

RULE. Prepare the given sum by the rule for New-England money, and divide by 4; the quotient is the answer in cents.

EXAMPLES.

1. Change £.461 to Federal money.

```
4)461000
  ------
  115250 cts.
```

Ans. 1152 dolls. 50 cts.

REDUCTION.

2. Change £.419 10s. 8½d. to Federal money.

    4)419535

    104883¾ cts.   Ans. 1048 dolls. 83¾ cts.

To change Federal money to New-York and North-Carolina currency.

RULE. As for Massachusetts currency, using 4 as a multiplier instead of 3; the value of a dollar being equal to four-tenths of a pound.

EXAMPLES.

1. Change 1684 dollars to New-York and North-Carolina currency.

    1684
       4

  Ans. £.673 12

2. Change 1048 dolls. 83¾ cents to New-York currency.

    1048,83¾
        4

    419,535
       20

    10,700
      12

    8,400
      4

    1,600   Ans. £.419 10s. 8¼d.

To change New-Jersey, Pennsylvania, Delaware and Maryland currency to Federal money, the dollar being 7s. 6d.

RULE. As the value of a dollar is equal to ⅜ of a pound, multiply the given sum, when it is pounds only, by 8, and divide by 3 for dollars. If there be shillings, &c. increase the sum in pence by ⅑ of the whole sum for cents.

EXAMPLES.

1. Change £.471 to Federal money.

    471
      8

  3)3768

  Ans. 1256 dollars.

REDUCTION. 39

2. Change £.480 19s. 9d. to Federal money.

    480 19 9
        20
    ―――
    9619
      12
    ―――
  9)115437
    ―――
   12826⅓

128263⅓ cents.   Ans. 1282 dolls. 63⅓ cts.

To change Federal money to New-Jersey, Pennsylvania, Delaware and Maryland currency.

RULE. Multiply the sum, when in dollars, by 3, and divide by 8 for pounds. If there be dollars and cents, multiply the given sum by 90, and the product (rejecting two figures on the right) is pence, or deducting $\frac{1}{10}$ of the sum gives the pence likewise.

EXAMPLES.

1. Change 1256 dollars to Pennsylvania currency.

    1256
       3
   ―――
  8)3768
  ―――
 Ans. £.471

2. Change 1282 dolls. 63⅓ cts. to Pennsylvania currency.

   128263⅓   Or $\frac{1}{10}$)128263⅓
      90   12826⅓
  ―――――   ―――――
12)115437,00   12)115437
 ―――――   ―――――
 20)9619—9   20)9619—9
 ―――――
Ans. £.480 19 9   £.480 19 9 as before.

To change South-Carolina and Georgia currency to Federal money, the dollar being 4s. 8d.

RULE. As the value of a dollar is equal to $\frac{7}{30}$ of a pound, if the sum be pounds only, multiply it by 30, and divide by 7 for dollars. If there be shillings, &c. annex two cyphers to the pence in the given sum, and divide by 56, the pence in a dollar, the quotient is the answer in cents.

REDUCTION.

EXAMPLES.
1. Change £.28 to Federal money.
28
30
─────
7)840
─────
120   Ans. 120 dolls.

2. Change £.11 4 8 to Federal money.
11 4 8
20
─────
224
12
─────
8×7=56.   8)269600
─────
7)33700
─────
4814⅔ cts.   Ans. 48 dols. 14⅔ cts.

To change Federal money to South-Carolina & Georgia currency.
RULE. Multiply the dollars by 7, and divide by 30 for pounds. If there be dollars and cents multiply by 56, and the product (rejecting two figures on the right) is the answer in pence.

EXAMPLES.
1. Change 540 dollars to S. Carolina and Georgia currency.
540
7
─────
3|0)378|0
─────
Ans. £.126

2. Change 48 dolls. 14⅔ cts. to South-Carolina currency.

4814⅔   56
56   2
─────   ─────
28884   7)112
24070   ─────
16   16
─────
12)2696,00
─────
20)224—8
─────
11 4 8   Ans. £.11 4 8

REDUCTION.

To change Canada and Nova-Scotia currency to Federal money, the dollar being 5 shillings.

RULE. As the value of a dollar is equal to one-fourth of a pound, multiply the sum, when in pounds, by 4, for dollars.

When there are shillings, &c. reduce the given sum to pence, annex two cyphers, and divide by 60, for cents.

EXAMPLES.

1. Change £.36 Canada currency to Federal money.

   36
   4

   Ans. 144 dolls.

2. Change £.528 12s. 6d. Canada currency to Federal money.

```
      20            Or thus,         528
    ─────                              4
    10572                           ─────
       12                            2112
    ─────                 10 shill. =   2
  6|0)126870|0             2s. 6d.  =   0 50
    ─────                           ─────
    211450 cts.                      2114 50
```
                    Ans. 2114 dolls. 50 cts.

To change Federal money to Canada and Nova-Scotia currency.

RULE. Divide the sum in dollars by 4 for pounds.

If there be dollars and cents, multiply the given sum by 60, and the product (rejecting two figures on the right) is the answer in pence.

EXAMPLES.

1. Change 144 dollars to Canada currency.

   4)144

   Ans. £.36

2. Change 2114 dols. 50 cts. to Canada or Nova-Scotia currency.

   211450
       60
   ─────────
   12)126870|00
   ─────────
   2|0)10572—6

   528 12 6   Ans. £.528 12s. 6d.

D 2

COMPOUND MULTIPLICATION

Is the multiplying of numbers of different denominations, by a simple figure or figures whose product shall be equal to a proposed number.

I. When the quantity does not exceed 12, multiply the price by the quantity, and the product will be the answer.

```
Multiply £.191  17   8½        £.913   11   9¾
    by              2.                     5.
                 ─────────             ─────────
  Ans. £.383   15   5.          £.4567  19   0¾
                 ─────────             ─────────

        £.980   19  11¾         £.209   18   4½
                    12                       9
              ─────────              ─────────
```

1. What will 7 yards of shalloon come to at 3s. 5d. per yard?

```
                    s.   d.
                    3    5.
                         7
                 ──────────
              £.1    3   11.
```

		s.	d.	£.	s.	d.
2.	4 lb. tea	6	8	1	6	8
3.	5 bushels rye	5	9	1	8	9
4.	6 gallons wine	7	5	2	4	6
5.	7 quintals fish	19	6	6	16	6
6.	9 cwt. iron	29	10	13	8	6
7.	11 gallons brandy	8	5	4	12	7
8.	12 quintals fish	22	10	13	14	0

II. If the number or quantity exceeds 12, and is to be found in the table, multiply by its component parts.

EXAMPLES.

```
                         s.   d.
   1. 14 yards durant at  2    5.
                               2
                        ─────────
                          4   10
                               7
                        ─────────
              Ans. £.1   13   10.
```

COMPOUND MULTIPLICATION. 43

		s.	d.		£.	s.	d.
2.	16 yards silk··at····	4	9		3	16	0
3.	20 lb. coffee········	1	9½		1	15	10
4.	28 gallons rum·······	6	5¾		9	1	5
5.	45 cwt. iron ········	29	6		66	7	6
6.	56 yards broadcloth	28	7		80	0	8
7.	63 pair shoes········	9	3		29	2	9
8.	84 quintals fish·····	18	6		77	14	0
9.	100 galls. molasses··	3	5½		17	5	10
10.	121 bushels corn·····	4	3		25	14	3
11.	144 gallons brandy ··	5	7¾		40	13	0

To multiply by fractional parts, as ½, ¾, ⅛, &c.

RULE. Multiply the price by the upper figure of the fraction, and divide the product by the lower, the quotient will be the answer; but when the upper figure is not more than one, dividing the price or sum by the lower figure gives the answer.

EXAMPLES.

1. What is ⅜ of a yard of cambric worth, at 12s. 6d. per yard?

```
    12  6
        3
   ─────────
   8)37  6
   ─────────
   Ans. 4s. 8¼d.
```

2. What is ¾ of a yard of broadcloth worth, at 35s. per yard?

```
    35            Or thus,  2)35
     3                      ─────────
   ─────                    2)17  6  price of half a yard.
   4)105                       8  9  a quarter.
   ─────                       ─────────
   Ans. 26s. 3d.              26  3
```

3. One quarter of a yard of fine linen, at 7s. 6d. per yard.

```
   4)7  6
   ─────────
   Ans. 1s. 10½d.
```

4. Multiply £4 5s. 3d. by ⅓, or take ⅓ of it.

```
   3)4  5  3
   ─────────
   Ans. £1  8  5
```

5. Multiply £.9 6*s.* 8*d.* by ⅞, or take ⅞ of it.

```
          9   6   8
                  7
       ------------
       8)65   6   8
       ------------
    Ans.£.8   3   4
```

III. When the number does not exceed the table, and it cannot be found in it, find the nearest to it, either less or greater; then, after having found the price of this number, add or subtract the value of so many, as it is less or greater than the given number.

EXAMPLES.

1. 37 bushels of corn, at 4*s.* 11*d.* per bushel.

```
             4  11
                 6
        ----------
         1   9   6
                 6
        ----------
         8  17   0  price of 36 bushels.
             4  11  price of  1 bushel.
        ----------            ----
Ans. £.9     1  11  price of 37 bushels.
```

			s.	*d.*		£.	*s.*	*d.*
2.	17¼	yards shalloon ·· at ··	2	8	Ans.	2	6	0
3.	23¾	lb. coffee ··········	1	10½	········	2	4	6¼
4.	57½	galls. rum ········	4	2½	······	12	1	11¾
5.	87¾	yds. baize ········	2	1	········	9	2	9¾
6.	109	quintals fish ······	14	6	······	79	0	6
7.	137¼	gallons of molasses	3	8¼	······	25	6	1¼

IV. When the number is above the table, find the price of each figure as in the following—

COMPOUND MULTIPLICATION.

EXAMPLES.

1. 178 yards of muslin at 4s. 5d. per yard.

```
           4   5
              10
         ─────────
           2   4   2
              10
         ─────────
          22   1   8  price of 100 yards.
          15   9   2  price of  70
           1  15   4  price of   8
         ─────────
Ans. £.39   6   2  price of 178 yards.
```

2. 284½ gallons of molasses, at 3s. 9½d. per gallon.

```
           3    9½
               10
         ─────────
           1   17   11
               10
         ─────────
          18   19    2
                     2
         ─────────
          37   18    4  price of 200 gallons.
          15    3    4  price of  80
               15    2  price of   4
           1   10¾   price of   ½
         ─────────
Ans. £.53  18    8¾ price of 284½ gallons.
```

		s.	d.		£.	s.	d.
3.	183 galls. gin····at····	7	5	····Ans.···	67	17	3
4.	345 quintals of fish····	23	9	········	409	13	9
5.	769¾ lb. coffee········	1	10	·········	70	11	2½
6.	809½ yards baize······	2	1½	··········	86	0	2¼
7.	2375½ galls. of molasses··	3	5½	········	410	15	3¼

8. Three barrels of N. E. rum, containing 31, 32½, and 33¼ gallons, at 4s. 7½d. per gallon. - Ans. £.22 7 5¼.

9. Four hogsheads of molasses, containing 97½, 99½, 105¼, and 111¾ gallons, at 3s. 8¾d. per gallon, are delivered by A to B, to whom he owed 258 dolls. It is required to know the balance, and in whose favour it is? Ans. 4s. 1½d. in favour of B.

COMPOUND MULTIPLICATION.

When the amount of a cwt. is required at a certain rate per lb.
RULE. Find the price of one or two quarters, and multiply the product by the component parts of a cwt.

1. 1 cwt. of Flour, at 3d. per lb.

```
        3
        7
       ───
      1 9
        8
       ───
     14 0  price of two quarters.
        2
       ───
Ans. £.1 8 0  price of one cwt.
```

Or by inverting the question thus,

```
    9  4  the price of 112 lb. at 1d. per lb.
    3
   ────
£.1 8  0  the price of 112 lb. at 3d. per lb.
```

			d.	per lb.	£.	s.	d.
2. Two cwt.	Flour	2½	per lb.	2	6	8
3. Three ··	Rice	2¾		3	17	0
4. Four ··	Iron	3¼		6	1	4
5. Five ··	Indigo 8s.1¼d.			250	16	8

1. What will 4000 feet of boards come to at 38s. 4d. per thousand?

```
        1 18 4
           4 M.
        ──────
Ans. £.7 13 4
```

2. 3,596 feet of boards at 36s. per thousand.

```
      3,596
         36
      ──────
      21576
      10788
      ──────
shills. 129,456

Ans. £.6 9 5.
```

In this example three figures are pointed off as a remainder, and the fourth figure of the product of this remainder, multiplied by 12, is set down for pence.

COMPOUND MULTIPLICATION.

3. 853 feet of boards at 30s. per thousand.

$$853$$
$$30$$
$$\overline{\text{shills. } 25{,}590}$$

Ans. £.1 5 7

4. 3,231 feet of 3 inch W. O. plank, 225s. £.36 6 11
5. 8,637 .... 2½ 150s. 64 15 6
6. ,960 .... 2 .. .. .. 100s. 4 16 0
7. ,888 .... 2½ pine, ... 100s. 4 8 9

Plank are sold per thousand of 2¼ inches, the usual thickness for planking vessels, and as there are generally other dimensions as 2 and 3 inches, the price of each is regulated by the price of the 2¼, adding to it, or subtracting from it, in such proportion as may be agreed on when purchasing. In the above example, taken from an actual sale, ½ of 150s. was added to it, for the three inch, and ⅓ deducted from it for the two inch, making the three inch 225s. and the two inch 100s. per thousand.

WEIGHTS AND MEASURES.

	lb.	oz.	dwt.	grs.		lb.	oz.	dwt.	grs.
Multiply	14	9	14	17		825	8	19	22
by				5					8
Product	74	0	13	13		6605	11	19	8

	T.	cwt.	qrs.	lb.		Cwt.	qr.	lb.	oz.	drs.
	19	17	3	25		17	1	14	11	14
				9						7

	T.	hhd.	gal.		T.	p.	hhd.	gal.
	87.	1.	57		28.	1	1	62
			5					7

What is the weight of 47 casks of rice, each weighing 2C. 1qr. 23lb. ? Ans. 115 cwt. 1 qr. 17 lb.

COMPOUND MULTIPLICATION.

BILLS OF PARCELS.

Boston, June 28, 1804.

Mr. GEORGE ROWE *bought of* WILLIAM RUSSELL,

		s.	d.			
8 pair worsted hose····at····	4	6	········	£.1	16	0
5 do. thread do.···········	3	2	··········	0	15	10
3 yards kerseymere ········	14	0	···········	2	2	0
6 do. muslin ···········	4	2	···········	1	5	0
2 do. tammy ············	1	8	············	0	3	4
4 shawls ················	7	6	··········	1	10	0

£.7 12 2

25 dols. 36 cts.

Portsmouth, 19th May, 1804.

Mr. THOMAS BARRINGTON
Bought of SIMON WILSON,

1¾ lb. Tea ·············4s.6················£.0 7 10½
4½ bushels corn··········5s.4···················
5 quarts brandy ······ 8s.4 per gallon ······
6 do. rum ········· 7s.6 do. ······
7¼ yards chintz ·········2s.5·················

£.3 11 0¾

11 dolls. 84⅓ cts.

Salem, 23d May, 1804.

Mr. AMOS GILES
Bought of LEMUEL KING,

10 boys' coloured hats, No. 1, at 4s.6··········£.2 5 0
12 ········do.········· 2, ·· 5s. ············
4 ········do.········· 3, ·· 5s.6···········
4 ········do.········· 9, ··10s. ···········
4 ········do.·········10, ··11s. ············
6 ········do.·········11, ··12s. ············
6 men's plain black do. 12, ··14s. ············

£.18 7 0
Trunk···· 1 4 0

£.19 11 0

65 dolls. 16⅔ cts.

COMPOUND DIVISION.

Mr. Nathan Perkins   Boston, 10th August, 1803.
                 Bought of George Allen,

64½ yds. striped nankins ··at·· 2s. ······ £.6 9 0
32 ells mode ··············· 3s. ·······
28½ yds. calico ············· 2s.4·······
2 groce gilt coat buttons···· 18s.6······
3 pieces russel ··············34s. ·······

                                 £.21 10 6.

                               71 dols. 75 cts.

Mr. William Sands   Newburyport, Sept. 10, 1803.
                 Bought of Stephen Nowlan,

2 pieces muslin ··············30s. ········£.3 0 0
25 yards Irish linen ··········· 2s. ·········
28¼ do. stormount calico······ 2s.6·········
28½ do. ·· red ···· do. ···· 2s.2·········
1 piece durant·················56s. ·········
2 pieces blue shalloon·········57s.6·········
50½ yards dimity ··············· 2s.6·········
3 pieces persian ·············· 84s. ·········

                               £.39 12 3

                               132 dols. 4 cts.

Received payment by his note of the above date, at three months.   *For Stephen Nowlan,*

                             Abraham Trusty.

······

COMPOUND DIVISION

Teacheth to find how often one number is contained in another of different denominations.

EXAMPLES.

1. Divide £.19 14s. 9½d. by 2.
        2)19 14 9½

     Ans. £.9 17 4¾

2. Divide £.900 11 9¾, by 3.   Ans. £.300 3 11¼
                   Prove this answer to be right.

E

COMPOUND DIVISION.

3. Divide £.121 7s. 9¾d. by 5.   Ans. £.24 5s. 6¾d.
4. Divide £.248 9s. 1½d. by 9.   Ans. £.27 12s. 1½d.
5. Divide £.1057 1s. 3d. by 12. Ans. £.88 1s. 9¼d.

II. If the divisor exceeds 12, and it be found in the table, divide by its component parts.

EXAMPLES.

1. Divide £.278 8s. 9d. between 45 men equally.

```
5)278  8  9
  ─────────
9) 55 13  9
```

Ans. £.6 3 9 each.

2. If 20 lb. of indigo cost £.7 5s. 10d. what is it per lb.?   Ans. 7s. 3½d.
3. If 24 yards of durant cost 62s. 6d. what is it per yard?   Ans. 2s. 7¼d.
4. If 72 bushels of corn cost £.20 9s. 6d. what is it per bushel?   Ans. 5s. 8¼d.
5. If 108 lb. of tea cost £.45 13s. 6d. what is one pound worth?   Ans. 8s. 5½d.
6. When £.166 13s. 4d. is paid for 500 gallons of rum, what is it per gallon?   Ans. 6s. 8d.
7. If 1000 gallons of molasses cost £.209 7s. 6d. what is it per gallon?   Ans. 4s. 2¼d.

III. If the divisor cannot be found by the multiplication of small numbers, as the preceding examples, divide by it as in the following   EXAMPLES.

1. Divide £.46 1s. 11d. by 37.

```
         £.  s.  d.
37)46  1  11(1  4  11 Ans.
   37
   ──
    9
   20
   ──
37)181(4
   148
   ───
    33
    12
   ──
37)407(11
   37
   ──
    37
    37
```

2. Divide £.33 13s. 8½d. by 23. Ans. £.1 9 3½.

3. If 345 quintals of fish cost £.409 13s. 9d. how much is it per quintal? Ans. 23s. 9d.

Dividing by fractional parts, as ½, ⅔, ⅘, &c. is the same as multiplying by them. See the Rule under Case II. in Compound Multiplication.

1. How much is ¾ of £.91 11s. 3d.

```
   91  11   3      Or thus  2)91  11   3
             3               ----------
  -------------              2)45  15   7½ one half the sum.
 4)274  13   9                 22  17   9¾ one quarter.
  -------------              -------------
Ans. £.68  13   5¼           £.68  13   5¼ answer.
```

2. Divide £.126 19s. 5¾d. by ⅘. Ans. £.101 11 7

3. If the whole of a ship is worth £.960 what is ⅝ worth? Ans. £.600

4. If ⅝ of a ship was sold for £.1056 2s. 1d. what was the whole valued at? Ans. £.1689 15 4

IV. Having the price of a hundred weight, to know how much it is per pound.

Rule. Find the price of 1 or 2 quarters, and then divide by the component parts.

1. If 1 cwt. of steel cost £.4. 6s. 4d. what is it per lb.?

```
 4)4  6  4              Or thus  2)4  6  4
 ---------                       ---------
 4)1  1  7  price of 1 qr.       7)2  3  2  price of 2 quarters.
 ---------                       ---------
 7)0  5  4¼                      8)0  6  2
 ---------                       ---------
Ans. 0  0  9¼ per lb.             0  0  9¼ per lb.
```

2. If 1 cwt. of flour cost 23s. 4d. what is it per lb.? Ans. 2½d.

3. When 2 cwt. of sugar cost £.8 17s. 4d. what is it per lb.? Ans. 9½d.

4. If 5 cwt. of iron cost £.8 15s. 0d. how much is it per lb.? Ans. 3¾d.

---

1. A mate and 3 seamen have to receive 600 dollars, for recapturing their vessel, of which the mate is to have two shares, and each seaman one share; how much is the part of each? Ans.—The mate's part is 240 dols. and each seaman's 120.

2. Capt. M. of the Jason, meets at sea with the wreck of the Hawk, of Boston, from which he takes sundry articles, which sell for 521 dollars 64 cents : two-thirds of this sum is awarded to the owners of the Hawk ; of the other ½ the owners of the Jason are to have ½, and the remainder is to be divided between the captain, mate, and nine seamen, allowing the captain 3 shares, the mate 2, and the seamen 1 share each; what is the respective part of those concerned ?

	dols.	cts.
Ans.—The owners of the Hawk	347	76
owners of the Jason	86	94
captain	18	63
mate	12	42
each seaman	6	21

DECIMAL FRACTIONS.

A DECIMAL FRACTION is that, whose denominator is an unit, with as many cyphers annexed to it as the numerator has places, and is usually expressed by writing the numerator only, with a point before it, called the separatrix ; thus, $\frac{5}{10}$, $\frac{25}{100}$, $\frac{125}{1000}$, are decimal fractions, and are expressed by ,5 ,25 ,125 respectively.

The figures to the left hand of the separatrix are whole numbers ; thus 4,5 yards is 4 yards and 5 tenths, or one half of another yard.

Cyphers placed to the right hand of decimals, make no alteration in their value ; for ,5 ,50 ,500 &c. are decimals of the same value, being each equal to ½ ; but when placed to the left hand, the value of the fraction is decreased in a tenfold proportion ; thus ,5 ,05 ,005 &c. are 5 tenth parts, 5 hundredth parts, 5 thousandth parts, respectively.

DECIMAL FRACTIONS.

The different value of figures will appear plainer by the following

TABLE.

```
          INTEGERS.        DECIMALS.
                 2,
                2 ,2
               2 0 ,0 2
              2 0 0 ,0 0 2
             2 0 0 0 ,0 0 0 2
            2 0 0 0 0 ,0 0 0 0 2
           2 0 0 0 0 0 ,0 0 0 0 0 2
          2 0 0 0 0 0 0 ,0 0 0 0 0 0 2
         2 0 0 0 0 0 0 0 ,0 0 0 0 0 0 0 2
```

(column labels: hundreds of millions. tens of millions. millions. hundreds of thousands. tens of thousands. thousands. hundreds. tens. units. tenths. hundredths. thousandths. ten thousandths. hundred thousandths. millionths. ten millionths. hundred millionths.)

From this table it appears, that as whole numbers increase in a tenfold proportion from units to the left hand, so decimals decrease in the same proportion to the right,—and that in decimals, as in whole numbers, the place of a figure determines its relative value.

ADDITION OF DECIMALS.

RULE. Place the given numbers so that the decimal points may stand directly under each other, then add as in whole numbers, and point off so many places for decimals to the right as are equal to the greatest number of the decimal places in any of the given numbers.

EXAMPLES.

263,51	42,23	2,1
149,28	18,47	,5
293,53	9,3	26,17
184,59	52,384	,7
129,4	2,1	5,
1020,31	124,484	34,47

E 2

Required the sum of twenty-nine and three tenths, three hundred and seventy-four and nine millionths, ninety-seven and two hundred and fifty-three thousandths, three hundred and fifteen and four hundredths, twenty-seven, one hundred and four tenths. Ans. 942,993009.

Required the sum of ten dollars and twenty-nine cents, ninety-three cents and three mills, nine cents and six mills, and two dollars and eight mills. Ans. 13 dols. 32 cts. 7 mills.

SUBTRACTION OF DECIMALS.

RULE. Place the given numbers so that the decimal points may stand directly under each other, and then point off the decimal places as in addition.

EXAMPLES.

From	219,42	87,26	57	311
Take	184,38	19,4	9,375	11,11
	35,04	67,86	47,625	299,89

From two thousand and sixteen hundredths take one thousand and four, and four millionths. Ans. 996,159996.

From twenty-four thousand nine hundred and nine and one tenth take fourteen thousand and twenty-nine thousandths. Ans. 10909,071.

Take eighty-five and seven hundred and thirty-seven thousandths from one hundred. Ans. 14,263.

From five hundred and thirty-one dollars two cents take one hundred and seventeen dollars three cents and four mills. Ans. 413 dols. 98 cts. 6 m.

MULTIPLICATION OF DECIMALS.

Multiply exactly as in whole numbers, and from the product cut off as many figures for decimals to the right hand as there are decimals in both the factors, but if the product should not have so many, supply the defect by prefixing cyphers.

DECIMAL FRACTIONS.

EXAMPLES.

Multiply	36,5	29,831	3,92
by	7,27	,952	196

	2555	59662	2352
	730	149155	3528
	2555	268479	392

Product	265,355	28,399112	768,32

Multiply	,285	,285	,29	124
by	,8	,003	,1	,06

Product	,2280	,000855	,029	7,44

NOTE. To multiply decimal fractions by 10, 100, 1000, &c. is only to remove the separatrix so many places towards the right as there are cyphers.

Thus, 7,362937 multiplied by { 10, 100, 1000, 10000 } is { 73,62937; 736,2937; 7362,937; 73629,37 }

Multiply twenty-nine and three tenths by seventeen.
Ans. 498,1.

Multiply twenty-seven thousandths by four hundredths.
Ans. ,00108.

Multiply two thousand and four and two tenths by twenty-seven.
Ans. 54113,4.

PRACTICAL QUESTIONS.

1. How much will 93 yards of shalloon come to at 53 cents per yard?

```
    93
   ,53
  ----
   279
   465
  ----
  49,29
```
Ans. 49 dolls. 29 cents.

2. At 21 cents 9 mills per lb. what will 187 lb. of coffee come to? Ans. 40 dols. 95 cents 3 mills.

DECIMAL FRACTIONS.

3. What will 27 cwt. of iron come to at 4 dollars 56 cents per cwt. ?   Ans. 123 dols. 12 cents.

4. How much will 281 yards of tape come to at 9 mills per yard ?   Ans. 2 dols. 52 cents 9 mills.

5. What will 371 yards of broadcloth come to at 5 dols. 79 cents per yard ?   Ans. 2148 dols. 9 cents.

6. How much will 29½ yards of mode come to at 75 cents per yard ?   Ans. 22 dols. 12 cents 5 mills.

7. What will 23,625 feet of boards come to at 8 dollars 25 cents per M. ?

```
   23,625
    8,25
   ──────
   118125
   47250
  189000
  ──────
  194,90625      Ans. 194 dols. 90 cents 6 mills.
```

8. How much will 712 feet of boards come to at 14 dollars per thousand ?   Ans. 9 dols. 96 cents 8 mills.

9. What will 25,650 feet of clear boards come to at 17 dols. 50 cents per thousand ?   Ans. 448 dols. 87 cents 5 mills.

			Dols.	Cts.		Dols.	Cts.	M.
10.	15,859	feet clear boards	17	50 per M.		277	53	2
11.	812	do.	14			11	36	8
12.	376	do.	12	75		4	79	4
13.	31,496	merchantable do.	8			251	96	8
14.	269	do.	6	75		1	81	5
15.	4,114	refuse do.	3	37		13	86	4
16.	393	maple do.		8 per foot		31	44	
17.	57	mahogany		32 do.		18	24	
18.	195	gallons molasses		57 per gall.		111	15	
19.	189	do. rum		93		175	77	
20.	243	yards baize		23 per yard		55	89	
21.	197	feet clear boards		2 per foot		3	94	

DIVISION OF DECIMALS.

RULE. Divide as in whole numbers, and from the right hand of the quotient point off as many places for decimals as the decimal places in the dividend exceed those of the divisor. If the places of the quotient are not so many as the rule requires, supply the defect by prefixing cyphers. If at any time there be a remainder, or the decimal places in the divisor are

DECIMAL FRACTIONS.

more than those in the dividend, cyphers may be annexed to the dividend, and the quotient carried to any degree of exactness.

EXAMPLES.

```
92),863972(,009391        ,853)89,000  (104,337, &c.
   828                       853
   ---                       ---
   359                       3700
   276                       3412
   ---                       ----
   837                       2880
   828                       2559
   ---                       ----
    92                       3210
    92                       2559
    --                       ----
                             6510
                             5971
                             ----
                              539
```

The various kinds that ever occur in division are included in the following cases, viz.

Divide	by	Ans.
,803	,22	3,65
,803	2,2	,365
,803	22	,0365
80,3	,22	365
80,3	2,2	36,5
80,3	22	3,65
222	,365	608,21+
222	3,65	60,821+
222	365	,60821+

As multiplying by 10, 100, 1000, &c. is only removing the separating point of the multiplicand so many places to the right hand as there are cyphers in the multiplier, so to divide by the same, is only removing the separatrix, in the same manner, to the left.

PRACTICAL QUESTIONS.

1. When butter is sold at 12 cents 8 mills per lb. how many lb. may be bought for 224 dollars?

```
,128)224,000(1750
    128
    ---
    960
    896
    ---
    640
    640
```
Ans. 1750 lb.

Here the cyphers annexed to the dividend being equal to the decimal places in the divisor, the quotient is a whole number.

2. If 673 bushels of wheat cost 786 dols. 73 cents 7 mills, what is it per bushel?

```
673)786,737(1,169
    673
    ---
    1137
     673
    ---
    4643
    4038
    ---
    6057
    6057
```
Ans. 1 dol. 16 cts. 9 mills.

In this example, as the divisor is a whole number, three places are pointed off in the quotient, to equal those in the dividend.

3. If 493 yards cost 4 dols. 43 cents 7 mills, what is it per yard? Ans. 9 mills.

4. If 125 gallons of molasses cost 95 dollars, what is 1 gallon worth? Ans. 76 cents.

5. If 205 yards of durant cost 107 dollars 62½ cents, what is it per yard? Ans. 52½ cents.

DECIMAL FRACTIONS.

REDUCTION OF DECIMALS.

Case I.

To reduce a vulgar fraction to its equivalent decimal.

RULE. Divide the numerator by the denominator, and the quotient will be the decimal required.

EXAMPLES.

1. Reduce ¾ to a decimal.

    4)3,00

    Ans. ,75

2. What is the decimal of ½ ?   Ans. ,5
3. What is the decimal of ¼ ?   Ans. ,25
4. What is the decimal of $\frac{3}{20}$?   Ans. ,15
5. What is the decimal of $\frac{17}{25}$?   Ans. ,68
6. Express ⅞ decimally.   Ans. ,875

Case II.

To reduce numbers of different denominations to their equivalent decimal values.

RULE. 1. Write the given numbers perpendicularly under one another for dividends, proceeding orderly from the least to the greatest.

2. Opposite to each dividend, on the left hand, place such a number for a divisor as will bring it to the next superior name, and draw a line between them.

3. Begin with the highest, and write the quotient of each division, as decimal parts, on the right hand of the dividend next below it, and the last quotient will be the decimal sought.

EXAMPLES.

1. Reduce 14s. 5½d. to the decimal of a pound.

4	2
12	5,5
20	14,4583

    Ans. ,7229

2. Reduce 15 shillings to the decimal of a pound. Ans. ,75
3. Reduce 3 qrs. 18lb. to the decimal of a cwt.
    Ans. ,910714+
4. Reduce 2 qrs. 2 nails to the decimal of a yard. Ans. ,625
5. Reduce 14 gals. 3 quarts to the decimal of a hogshead.
    Ans. ,2341+

DECIMAL FRACTIONS.

Case III.
To find the decimal of any number of shillings, pence and farthings, by inspection.

RULE. Write half the greatest even number of shillings for the first decimal figure, and let the farthings, in the given pence and farthings, possess the second and third places; observing to increase the second place by 5, if the shillings be odd, and the third place by 1, when the farthings exceed 12, and by 2 when they exceed 37.

EXAMPLES.

1. Find the decimal of 13s. 9¾d. by inspection.

,6   half of 12s.
5   for the odd shilling.
39 farthings in 9¾d.
2   for excess of 37
_____
,691

2. Find by inspection the decimal of 15s. 8¼d. 9s. 3½d. 19s. 6¾d. 3s. 6d. and 2s. 11½d.   Ans. ,784 ,465 ,978 ,175 ,148.

Case IV.
To find the value of any given decimal in the terms of the integer.

RULE. 1. Multiply the decimal by the number of parts in the next less denomination, and cut off as many places for the remainder to the right hand as there are places in the given decimal.

2. Multiply the remainder by the parts in the next inferior denomination, and cut off a remainder as before.

3. Proceed in this manner through all the parts of the integer, and the several denominations, standing on the left hand make the answer.

EXAMPLES.

1. Find the value of ,691 of a pound.

,691
20
_____
13,820
12
_____
9,840
4
_____
3,360   Ans. 13s. 9¾d.

2. What is the value of ,9 of a shilling?   Ans. 10¾d.
3. What is the value of ,592 of a cwt.?
   Ans. 2 qrs. 10 lb. 4 oz. 13 + drs.
4. What is the value of ,258 of a tun of wine?
   Ans. 1 hhd. 2 + galls.
5. What is the value of ,12785 of a year?
   Ans. 46 days 15 hours 57 minutes 57 + sec.

DECIMAL TABLES OF COIN, WEIGHT AND MEASURE.

TABLE I.

ENGLISH COIN.

1*l.* the Integer.

Sh.	dec.	Sh.	dec.
19	,95	9	,45
18	,9	8	,4
17	,85	7	,35
16	,8	6	,3
15	,75	5	,25
14	,7	4	,2
13	,65	3	,15
12	,6	2	,1
11	,55	1	,05
10	,5		

Pence.	Decimals.
6	,025
5	,020833
4	,016666
3	,0125
2	,008333
1	,004166

Farth.	Decimals.
3	,003125
2	,0020833
1	,0010416

TABLE II.

ENG. COIN.   1 Shill.

LONG MEAS.   1 Foot.

The Integer.

Pence and Inches.	Decimals.
6	,5
5	,416666
4	,333333
3	,25
2	,166666
1	,083333

Farth.	Decimals.
3	,0625
2	,041666
1	,020833

TABLE III.

TROY WEIGHT.

1 *lb.* the Integer.

Ounces the same as Pence in the last Table.

Penny weight.	Decimals.
10	,041666
9	,0375
8	,033333
7	,029166
6	,025
5	,020833
4	,016666
3	,0125
2	,008333
1	,004166

Grains.	Decimals.
12	,002083
11	,001910
10	,001736
9	,001562
8	,001389
7	,001215
6	,001042
5	,000868
4	,000694
3	,000521
2	,000347
1	,000173

1 *oz.* the Integer.

Pennyweight the same as Shillings in the first Table.

Grains.	Decimals.
12	,025
11	,022916
10	,020833
9	,01875
8	,016666
7	,014583

Grains.	Decimals.
6	,0125
5	,010416
4	,008333
3	,00625
2	,004166
1	,002083

TABLE IV.

AVOIRDUPOIS WT.

112 *lb.* the Integer.

Qrs.	Decimals.
3	,75
2	,5
1	,25

Pounds	Decimals.
14	,125
13	,116071
12	,107143
11	,098214
10	,089286
9	,080357
8	,071428
7	,0625
6	,053571
5	,044643
4	,035714
3	,026786
2	,017857
1	,008928

Ounces	Decimals.
8	,004464
7	,003906
6	,003348
5	,002790
4	,002232
3	,001674
2	,001116
1	,000558

¼ oz.	Decimals.
3	,000418
2	,000279
1	,000139

F

DECIMAL TABLES OF COIN, WEIGHT AND MEASURE.

TABLE V.

AVOIRDUPOIS WT.

1 lb. the Integer.

Oz.	Decimals.
8	,5
7	,4375
6	,375
5	,3125
4	,25
3	,1875
2	,115
1	,0625

Drm.	Decimals.
8	,03125
7	,027343
6	,023437
5	,019531
4	,015625
3	,011718
2	,007812
1	,003906

TABLE VI.

LIQUID MEASURE.

1 Tun the Integer.

Gals.	Decimals.
100	,396825
90	,357141
80	,317460
70	,27
60	,238095
50	,198412
40	,158730
30	,119047
20	,079365
10	,039682
9	,035714
8	,031746
7	,027
6	,023809

Gals.	Decimals.
5	,019841
4	,015873
3	,011904
2	,007936
1	,003968

Pints.	Decimals.
4	,001984
3	,001488
2	,000992
1	,000496

A hogshead the Integer.

Gals.	Decimals.
30	,476190
20	,317460
10	,158730
9	,142857
8	,126984
7	,111111
6	,095238
5	,079365
4	,063492
3	,047619
2	,031746
1	,015873

Pints.	Decimals.
3	,005952
2	,003968
1	,001684

TABLE VII.

MEASURE.

Liquid.   Dry.

1 Gallon.   1 Quarter.

Integer.

Pt.	Decim.	Bu.
4	,5	4
3	,375	3
2	,25	2
1	,125	1

Q.pt.	Decim.	Pk.
3	,09375	3
2	,0625	2
1	,03125	1

Decimals.	Q pks.
,0234375	3
,015625	2
,0078125	1

Decimals.	Pts.
,005859	3
,003906	2
,001953	1

TABLE VIII.

LONG MEASURE.

1 Mile the Integer.

Yards.	Decimals.
1000	,568182
900	,511364
800	,454545
700	,397727
600	,340909
500	,284091
400	,227272
300	,170454
200	,113636
100	,056818
90	,051136
80	,045454
70	,039773
60	,034091
50	,028409
40	,022727
30	,017045
20	,011364
10	,005682
9	,005114

Decimal Tables of COIN, WEIGHT and MEASURE.

Yards.	Decimals.
8	,004545
7	,003977
6	,003409
5	,002841
4	,002273
3	,001704
2	,001139
1	,000568

Feet.	Decimals.
2	,0003787
1	,0001894

Inches.	Decimals.
6	,0000947
5	,0000789
4	,0000631
3	,0000474
2	,0000319
1	,0000158

TABLE IX.

Time.

1 Year the Integer.

Months the same as Pence in the second Table.

Days.	Decimals.
365	1,000000
300	,821918
200	,547945
100	,273973
90	,246575
80	,219178
70	,191781
60	,164383
50	,136986
40	,109589
30	,082192
20	,054794
10	,027397
9	,024657

Days.	Decimals.
8	,021918
7	,019178
6	,016438
5	,013698
4	,010959
3	,008219
2	,005479
1	,002739

1 Day the Integer.

Hours.	Decimals.
12	,5
11	,458333
10	,416666
9	,375
8	,333333
7	,291666
6	,25
5	,208333
4	,166666
3	,125
2	,083333
1	,041666

Minutes.	Decimals.
30	,020833
20	,013888
10	,006944
9	,00625
8	,005555
7	,004861
6	,004166
5	,003472
4	,002777
3	,002083
2	,001388
1	,000694

TABLE X.

Cloth Measure.

1 Yard the Integer.

Quarters the same as Table IV.

Nails.	Decimals.
2	,125
1	,0625

TABLE XI.

Lead Weight.

1 Fother the Integer.

Hund.	Decimals.
10	,512820
9	,461538
8	,410256
7	,358974
6	,307692
5	,256410
4	,205128
3	,153846
2	,102564
1	,051282

Qrs.	Decimals.
2	,025641
1	,012820

Pounds.	Decimals.
14	,0064102
13	,0059523
12	,0054945
11	,0050366
10	,0045787
9	,0041208
8	,0036629
7	,0032051
6	,0027472
5	,0022893
4	,0018315
3	,0013736
2	,0009157
1	,0004578

The Single Rule of Three Direct.

THE Single Rule of Three Direct teaches, from three numbers given, to find a fourth, that shall be in the same proportion to the third as the second is to the first.

If *more* requires *more*, or *less* requires *less*, the proportion is direct.

RULE 1. Make the number that is the demand of the question, the third term, the number that is of the same name or quality, the first term, and the remaining number will be the middle term.

Reduce the first and third numbers into the same, and the second into the lowest denomination mentioned.

2. Multiply the second and third numbers together, and divide the product by the first, and the quotient (if there be no remainder) is the answer, or fourth number required.

If, after division there be a remainder, reduce it to the next denomination below that to which the second number was reduced, and divide by the same divisor as before, and the quotient will be of this last denomination. Proceed thus with all the remainders till you have reduced them to the lowest denomination, which the second number admits of, and the several quotients taken together will be the answer required.

The method of proof is by reversing the question.

EXAMPLES.

1. If 2 yards of cloth cost 4s. what will 125 yards come to?

```
   yds.  s.   yds.              yds.   £.  s.   yds.
If  2  : 4 :: 125       Proof if 125 : 12 10 :: 2
            4                          20
         ─────                        ─────
        2)500                          250
         ─────                          2
       20)250                        ─────
         ─────                     125)500(4 shillings.
   Ans.  £.12 10                      500
                                      ─────
```

SINGLE RULE OF THREE DIRECT.

2. If 1 bushel of corn cost 75 cents, what will 257 bushels come to?

```
    bush.   cts.    bush.
If    1  :   75  ::  257
                      75
                    ─────
                     1285
                     1799
                    ─────
                    192,75    Ans. 192 dols. 75 cts.
```

3. What will 931 yards of shalloon come to at 55 cts. 4 ms. per yard?   Ans. 515 dols. 77 cts. 4 ms.

4. How many bushels of wheat at 1 dol. 12 cts. per bushel can I have for 81 dols. 76 cts.?   Ans. 73 bushels.

5. What will 94 cwt. of iron come to at 4 dols. 97 cts. 2 ms. per cwt.?   Ans. 467 dols. 36 cts. 8 ms.

6. What will 349 lbs. of beef come to at 2d. per lb.?   Ans. £.2 18 2

7. At 3s. per yard what will 59 yards of cloth come to?   Ans. £.8 17 0

Prove this answer to be right.

8. How many lbs. of beef at 5 cts. per lb. may be bought for 29 dols. 85 cts.?

```
       cts.   lb.   dols. cts.
    If  5  :  1  ::  29,85
                         1.
                      ─────
                ,05)29,85
                      ─────
                      597      Ans. 597 lb.
```

9. How many hhds. of salt at 4 dols. 90 cts. per hhd. can I have for 392 dols.?   Ans. 80 hhds.

10. How many lbs. of coffee, at 1s. 7d. per lb. may be bought for £.8 12 7?   Ans. 109 lb.

SINGLE RULE OF THREE DIRECT.

11. When 25 yds. of cloth cost £.2 12 1, what is it per yd.?

```
     yd.    £.  s. d.   yd.
If   25  :  2  12  1  ::  1
            20
           ———
            52
            12
           ———
           625
             1
           ———
25)625(12 | 25
   50         ———
   ———         2s. 1d.
   125
   125
   ———
```
Ans. 2s. 1d.

12. If 56 bushels of corn cost 42 dols. 56 cts. what is it per bushel?

```
       bush.  dols.cts.  bush.
If      56  :  42,56  ::  1
                  1
              ————————
       56)42,56(,76
          392
          ————
           336
           336
          ————
```
Ans. 76 cts.

13. If 112 lbs. of beef cost 18s. 8d. what is it per lb.?
Ans. 2 pence.

14. If 673 bushels of rye cost 769 dols. 23 cts. 9 ms. what is 1 bushel worth?   Ans. 1 dol. 14 cts. 3 ms.

15. What is 1 yard of baize worth, when 97 yards cost £.10 12s. 2¼d.   Ans. 2s. 2¼d.

16. When iron is sold at 5 dols. 4 cts. per cwt. what is it per pound?   Ans. 4 cts. 5 ms.

17. If 891 gallons of molasses cost £.176 6s. 10½d. what is it per gallon?   Ans. 3s. 11½d.
Prove this answer to be right.

18. What will 253 quintals of fish come to, at 17s. 6d. per quintal?   Ans. £.221 7 6.

SINGLE RULE OF THREE DIRECT. 67

19. At 5 dols. 50 cts. per thousand, what will 37 thousand of boards come to? Ans. 203 dols. 50 cts.

20. What will 4 hhds. of rum come to, containing viz. 79½, 84, 101½, and 112 gals. at 6s. 9d. per gal.? Ans. £.127 4 9

21. What will 327 hhds. of salt come to, at 5 dols. 25 cts. per hhd.? Ans. 1716 dols. 75 cts.

22. If 3 and 4 make 9, how many will 6 and 8 make?
Ans. 18

23. If a chest of Hyson tea, weighing 79 lb. neat, cost £.32 11s. 9d. what is it per lb.? Ans. 8s. 3d.

24. B owes £.2119 17s. 6d. and he is worth but £.1324 18s. 5¼d.; if he delivers this to his creditors, how much do they receive on the pound? Ans. 12s. 6d.

25. A owes B £.569 6s. 8d. but failing in trade, he is able to pay but 15s. 6d. on the pound; how much is B to receive, and what is his loss? Ans.—He is to receive £.441 4 8
His loss is 128 2 0

26. A merchant failing in trade, owes in all 2947.5 dols. and delivers up his whole property, worth 21894 dols. 3 cts.; how much per cent. does he pay, and what is B's loss, to whom he owed 325 dols.? Ans.—He pays 74 dols. 28 cts. per cent.
And B loses 83 dols. 59 cts.

27. How much will 4 cwt. 1 qr. 19 lb. of butter come to, at 9d. per lb.?
```
              lb.
        400 = 4 hundred.
         48 = excess, 12 per cent.
         28 = 1 quarter.
         19
      lb.    d.
   If  1  :  9  ::  495
                     9
                   ─────
               12)4455
                   ─────
               20)371  3
                   ─────
```
Ans. £.18 11s. 3d.

28. If 3 qrs. 26 lb. of steel cost 13 dols. 20 cts. what is it per pound? Ans. 12 cents.

SINGLE RULE OF THREE DIRECT.

29. If 16 cwt. 3 qrs. of steel cost 157 dols. 45 cts. what is 1 qr. worth ?   Ans. 2 dols. 35 cts.
Prove this answer to be right.

30. A captain of a ship is provided with 18000 lb. of bread for 150 seamen, of which each man eats 4 lb. per week, how long will it last them ?   Ans. 30 weeks.

31. How long would 2295 lb. of beef last for 45 seamen, if they get 1 lb. each, and that three times a week ?
Ans. 17 weeks.

32. Suppose 120 seamen are provided with 7200 gallons of water for a cruise of 4 months, each month 30 days; how much is each man's share per day ?   Ans. 2 quarts.

33. A ship's company of 16 men is on an allowance of 6 ounces of bread per day, when meeting with a vessel from which they are supplied with 2 cwt. of bread, what addition will this make to their daily allowance, if they suppose their voyage to last 28 days ?   Ans. 8 ounces.

34. If 17 tuns 2 hhds. of wine cost 5468 dols. 40 cts. how much is one pint worth ?   Ans. 15 cts. 5 ms.

35. How much will 4 pieces of linen, containing, viz. $35\frac{1}{2}$, 36, $37\frac{1}{2}$, and 38 yards come to, at 79 cts. per yard ?
Ans. 116 dols. 13 cts.

36. How many crowns of 110 cents each will pay a debt of £.82 16s. 7d. ?   Ans. 251 crowns.

37. If 203 tons 9 cwt. 3 qrs. 3 lb. of tallow cost £.4558 3s. 0d. what does 1 ton cost ?   Ans. £.22 8 0

38. How many cwt. of rice may be bought for 487 dols. 50 cts. when 7 lb. cost 25 cents ?   Ans. 121 cwt. 3 qrs. 14 lb.

39. When 9 dols. 36 cts. is paid for 2 qrs. 22 lb. of sugar, what is 7 lb. worth ?   Ans. 84 cents.

40. When 47 cwt. 3 qrs. of sugar cost £.182 4s. 11d. what is 1 qr. worth ?   Ans. 19s. 1d.

41. If 6 lb. 6 oz. Avoirdupois cost 5 dols. 10 cts. what is it per ounce ?   Ans. 5 cents.

42. Bought 40 tubs of butter weighing 36 cwt. 2 qrs. 14 lb. neat, for 472 dols. 2 cts. ; paid cooperage 12 cts. per tub; salt and labour 4 dols. 83 cts. 8 mills ; storage 6 dols. 48 cts.— I would know what it stands me in per lb. ?   Ans. 11 cts. 9 ms.

SINGLE RULE OF THREE DIRECT. 69

43. How much will a grindstone, 32 inches diameter, and 6 inches thick, come to, at 5s. per cubic foot?

See Reduction,
cubic measure.

```
          32   the diameter.
          16 = half the diameter.
         ────
          48
          16
         ────
         288
          48
         ────
         768
           6
```

        inch.   *s.*   *s. d.*
If 1728 : 5 :: 4608 : 13 4 Ans. 13s. 4d.

44. What will a grindstone, 28 inches diameter, and 3½ inches thick, come to, at 1 dol. 90 cts. per cubic foot?
                                        Ans. 2 dols. 26 cts.

45. When a man's yearly income is 949 dollars, how much is it per day?   Ans. 2 dols. 60 cts.

46. At 4½ per cent. what is the commission on 1525 dols.?
                        Ans. 68 dols. 62 cts. 5 ms.

47. What is the interest of 456 dollars for 1 year, at 6 per cent.?   Ans. 27 dols. 36 cts.

48. At 5 dols. 50 cts. per M. what will 21,186 feet boards come to?   Ans. 116 dols. 52 cts. 3 ms.

49. When boards are sold at 18 dols. per M. what is it per foot?   Ans. 1 cent, 8 mills.

50. What will 98 feet of boards come to at 4 cts. per foot?
                        Ans. 3 dols. 92 cts.

51. What will 49 thousand 3 hundred and 25 casts of staves come to at 17 dols. per thousand?

NOTE. Staves are counted by casting three at a time; 40 casts make 1 hundred, and 10 hundred 1 thousand.

```
       M.    dols.      M.  h.  c.
  If   1  :  17  ::    49   3  25
      10                10
      ──                ───
      10               493
      40                40
      ──               ────
Casts 400             19745
                              dols. cts. m.
                         Ans. 839  16   2
```

52. What will 19 M. 8 and 15 casts of white oak hhd. staves come to, at 31 dols. per M.?   Ans. 614 dols. 96 cts. 2 ms.

SINGLE RULE OF THREE DIRECT.

53. What will 22 M. 9 and 37 casts of red oak hhd. staves come to, at 13 dols. per M.? Ans. 298 dols. 90 cts. 2 ms.

54. What will 56 bundles of hoops come to at 25 dols. per M. of 30 bundles?

NOTE. Hoops are sometimes bound in bundles of 30 hoops each, and 4 such bundles are 1 hundred, and 10 hundred or 40 bundles, 1 thousand. But they are generally bound in bundles of 40 each, 3 bundles making 1 hundred, and 10 hundred or 30 bundles, 1 thousand.

```
            hund. dols.    3)56              Or  bund. dols.  bund.
        If   10  :  25  ::  18⅔ hundreds        30 : 25  ::  56
                             25                               25
                            ———                              ———
                             90                              280
                             36                              112
                             16⅔                            ———
                            ———                           3|0)140|0
                         1|0)46|6⅔                          ———
                            ———                             46,66⅔
                            46,6⅔
```

Ans. 46 dols. 6⅔ dimes, or 66⅔ cts.

55. How many bushels of salt, at 4 dols. 75 cts. per hhd. can I have for 326 dollars?

dols.cts. bush.   dols.
If 4 75 : 8 :: 326   Ans. 549 bushels, when measured on board the vessel.

If 4 75 : 7½ :: 326   Ans. 514 bushels three pecks, nearly, when measured ashore.

56. What is the tax on lands, &c. valued at 2957 dols. in the direct tax, at 28 cents and 3 mills on the 100 dollars?
Ans. 8 dols. 36 cts. 8 ms.

57. What is the tax on a house, valued at 900 dollars, in the direct tax, at $\frac{3}{10}$ per cent.?

```
           dols     dols.    dols.
       If  100  :  ,3   ::   900
                             ,3
                            ———
                      100)270,0
```

Ans. 2 dols. 70 cts.

Or, As $\frac{3}{10}$ per cent. is equal to 3 mills on the dollar, multiplying the sum in dollars by 3, gives the answer in mills.

SINGLE RULE OF THREE DIRECT. 71

EXAMPLE.

58. What is the tax on 753 dollars at $\frac{3}{10}$ per cent.?

   753 dollars
     3 mills
   ―――――
   2259 mills. Ans. 2 dols. 25 cts. 9 ms.

59. Find the tax on the following sums—

  dols.   dols. cts.
viz: 1550 at $\frac{4}{10}$ per cent. .... Ans. 6. 20
  4560 $\frac{5}{10}$ 22 80
  7850 $\frac{6}{10}$ 47 10
 12680 $\frac{7}{10}$ 88 76
 16950 $\frac{8}{10}$ 135 60
 24620 $\frac{9}{10}$ 221 58
 35840 1 .... 358 40

60. What will a piece of land, measuring 48 feet in length and 40 feet in width at each end, amount to at 20 dollars per square rod?

        feet.
         48
         40
  feet. dols. ―――
If 272$\frac{1}{4}$: 20 :: 1920
  By decimals.   Ans. 141 dols. 4 cts.
If 272,25 : 20 :: 1920

61. A charter-party for a vessel of 186 tons commenced on 28th of May, and ended on the 10th of October following: What does the hire amount to for that time, at 2 dols. per ton per month of 30 days?

             days.
          May 4
          June 30
          July 31
          August 31
    186 tons.   September.. 30
    2 dols. per mo. October.... 10
 days. ―――   ―――
If 30 : 372   136
    · 136
    ―――
    2232
    1116
     372
   ―――――
 3,0)5059,2
   ―――――
   1686,40   Ans. 1686 dols. 40 cts.

In calculating the time, the days of receiving and discharging the vessel are both included.

INVERSE PROPORTION.

WHEREAS in the Rule of Three Direct, more requires more, and less requires less, in this rule more requires less and less requires more.

RULE. After stating the terms as in the Rule of Three Direct, multiply the first and second terms together, and divide the product by the third, and the quotient is the answer.

EXAMPLES.

1. If 100 workmen complete a piece of work in 12 days, how many are sufficient to do it in 3 days?

```
    d.      m.      d.
    12  :  100  ::  3
           12
          ────
       3)1200
          ────
           400           Ans. 400 men.
```

2. If 8 boarders drink a barrel of cider in 12 days, how long would it last if 4 more came among them? Ans. 8 days.

3. A ship's company of 15 persons is supposed to have bread to last their voyage, allowing each 8 ounces per day—when they pick up a crew of 5 persons in distress, to whom they are willing to communicate, what will the daily allowance of each person then be? Ans. 6 ounces.

4. When wheat is sold at 93 cts. per bushel, the penny loaf weighs 12 ounces—what must it weigh when the wheat is 1 dol. 24 cts. per bushel? Ans. 9 ounces.

5. How many yards of baize, 3 qrs. wide, will line a cloak, which has in it 12 yds. of camblet, half yard wide? Ans. 8 yds.

6. Suppose 400 men in a garrison are provided with provisions for 30 days, how many men must be sent out, if they would have the provisions last 50 days? Ans. 160 men.

7. What sum should be put to interest to gain as much in 1 month, as 127 dollars would gain in 12 months? Ans. 1524 dols.

COMPOUND PROPORTION.

Compound Proportion teaches to resolve such questions, as require two or more statings by simple proportion.

Rule. State the question, by placing the three conditional terms in this order: that which is the principal cause of gain, loss, or action, possesses the first place; that which denotes space of time or distance of place, the second; and that which is the gain, loss, or action, the third; then place the other two terms, which move the question, under those of the same name, and if the blank place fall under the third, multiply the three last terms for a dividend, and the two first for a divisor: but if the blank fall under the first or second place, multiply the first, second, and last terms together for a dividend, and the other two for a divisor; and the quotient will be the answer.

EXAMPLES.

1. If £.100 in 12 months gain £.5, how much will £.400 gain in 3 months?

```
    £.      mo.      £.
    100  :  12   ::  5
    400  :   3
   _____
    100    1200
     12       5
   _____
    12|00)60|00
     £.5              Ans. £.5
```

2. If 8 men make 24 rods of wall in 6 days, how many men will build 18 rods in 3 days?

```
    m.     d.      r.
    8   :  6   ::  24
    8          18
                6
          _____
           24    108
            3      8
          _____
          72 )864( 12
             72
            ___
            144
            144
            ___
                    Ans. 12 men.
```

COMPOUND PROPORTION.

3. If a family of 9 persons spend 450 dollars in 5 months, how much would be sufficient to maintain them 8 months, if five more were added to the family?   Ans. 1120 dolls.

4. What is the interest of £.240 for 50 days, at 5 per cent. per annum?

```
         £.      days.     £.
         100  :  365  : :  5
         240  :  50
          50
         ―――
    100  12000
    365      5
   ―――― ―――――
   365|00)600|00(1  12  10½
          365
          ―――
          235
           20
          ―――
     365)4700(12
         4380
         ――――
          320
           12
         ――――
     365)3840(10
         365
         ―――
          190
            4
         ―――
      365)760(2
          730
          ―――
           30              Ans. £.1  12  10½
```

N. B. By omitting to multiply by the rate per cent. and dividing 36500 by it, are found the fixed divisors of 7300 for 5, and 6083 for 6 per cent. per annum, sometimes used in calculating interest.

COMPOUND PROPORTION.

5. What is the interest of 654 dollars for 164 days, at 6 per cent. per annum?

```
      100                    654 dollars.
      365                    164
    _____                 _____
  6) 36500                   2616
    _____                   3924
      6083 the fixed divisor, 654
  found as above directed.  _____
                            6083) 107256 (17,632
                                  6083
                                  _____
                                  46426
                                  42581
                                  _____
                                  38450
                                  36498
                                  _____
                                  19520
                                  18249
                                  _____
                                  12710
                                  12166
                                  _____
                                   544   Ans. 17d. 63c. 2m.
```

6. What is the interest of 947 dollars, for 294 days, at 5 per cent. per annum?

```
                      947 dolls.
                      294
                      _____
                      3788
                      8523
                      1894
                      _____
  Fixed divisor  7300) 278418 (38,139
                      21900
                      _____
                      59418
                      58400
                      _____
                      10180
                       7300
                      _____
                      28800
                      21900
                      _____
                      69000
                      65700
                      _____
                       3300   Ans. 38 dols. 13c. 9m.
```

VULGAR FRACTIONS.

FRACTIONS, or broken numbers, are expressions for any assignable parts of an unit; and are represented by two numbers, placed one above the other, with a line drawn between them.

The number above the line is called the *numerator*, and that below the line the *denominator*.

The denominator shews how many parts the integer is divided into, and the numerator shews how many of those parts are meant by the fraction.

Fractions are either proper, improper, compound, or mixed.

1st. A *proper fraction* is when the numerator is less than the denominator, as $\frac{1}{3}$, $\frac{2}{5}$, $\frac{9}{11}$, $\frac{53}{86}$, &c.

2d. An *improper fraction* is when the numerator is either equal to or greater than the denominator, as $\frac{8}{8}$, $\frac{11}{9}$, $\frac{12}{2}$, $\frac{35}{20}$, &c.

3d. A *compound fraction* is a fraction of fractions, and known by the word *of*, as $\frac{1}{2}$ of $\frac{2}{3}$, $\frac{7}{8}$ of $\frac{9}{10}$, $\frac{15}{16}$ of $\frac{21}{28}$, &c.

4th. A *mixed number* or *fraction* is composed of a whole number and fraction, as $8\frac{2}{7}$, $17\frac{1}{2}$, $29\frac{3}{4}$, &c.

I. To reduce a simple fraction to its lowest terms.

RULE. Find a common measure by dividing the lower term by the upper, and that divisor by the remainder, continuing till nothing remains; the last divisor is the common measure; then divide both parts of the fraction by the common measure, the quotients express the fraction required.

NOTE. If the common measure happens to be 1, the fraction is already in its lowest term; and when a fraction hath cyphers at the right hand, it may be abbreviated by cutting them off, as $\frac{5}{5}|\frac{0}{0}$.

EXAMPLES.

1. Reduce $\frac{91}{117}$ to its lowest term.

```
91)117(1
    91
    ──
    26)91(3
       78
       ──
Common measure   13)26(2
                    26
                    ──
```

$13)\frac{91}{117}(\frac{7}{9}$ the answer.

VULGAR FRACTIONS. 77

Or; divide the terms of the fraction by any number that will divide them without a remainder.; divide the quotients in the same manner, and so on, till no number will divide them both, and the last quotients express the fraction in its lowest terms.

EXAMPLES.

2. Reduce $\frac{192}{576}$ to its lowest terms..

$$\underset{576}{192} \;\overset{(8)}{=}\; \underset{72}{24} \;\overset{(8)}{=}\; \underset{9}{3} \;\overset{(3)}{=}\; \frac{1}{3} \quad \text{the answer.}$$

3. Reduce $\frac{144}{168}$ to its lowest terms..   Ans. $\frac{6}{7}$.
4. Reduce $\frac{132}{168}$ to its lowest terms..   Ans. $\frac{2}{3}$.
5. Reduce $\frac{2619}{2231}$ to its lowest terms..   Ans. $1\frac{1}{9}$.

II. *To reduce a mixt number to an improper fraction.*

RULE. Multiply the whole numbers by the denominator of the fraction, and to the product add the numerator for a new numerator; and place it over the denominator.

NOTE. To express a whole number fraction-wise, set 1 for a denominator to the given number.

EXAMPLES.

1. Reduce $5\frac{3}{8}$ to an improper fraction..
$$5 \times 8 + 3 = \tfrac{43}{8} \text{ the answer..}$$

2. Reduce $183\frac{5}{21}$ to an improper fraction..   Ans. $\frac{3848}{21}$.
3. Reduce $27\frac{2}{5}$ to an improper fraction.   Ans. $\frac{245}{9}$..
4. Reduce $514\frac{5}{16}$ to an improper fraction..   Ans. $\frac{8229}{16}$..

III. *To reduce an improper fraction to its proper terms..*

RULE. Divide the upper term by the lower, and the quotient will be the whole number; the remainder, if any, will be the numerator to the fractional part..

EXAMPLES.

1. Reduce $\frac{17}{5}$ to its proper terms.
$$5)17(3\tfrac{2}{5} \text{ the answer..}$$
$$\underline{15}$$
$$2$$

2. Reduce $\frac{245}{9}$ to its proper terms..   Ans. $27\frac{2}{9}$..
3. Reduce $\frac{8229}{16}$ to its proper terms..   Ans. $514\frac{5}{16}$..

G 2.

IV. To find the least common multiple or denominator.

RULE. Divide the given denominators by any number that will divide two or more of them without a remainder, and set the quotients and the undivided numbers underneath. Divide these quotients and undivided numbers by any number that will divide two or more of them as before, and thus continue, till no two numbers are left capable of being lessened.

Multiply the last quotients and the divisor or divisors together, and the product will be the least common denominator required.

EXAMPLES.

1. What is the least common measure of $\frac{5}{9}, \frac{7}{8}, \frac{6}{15},$ & $\frac{3}{16}$?

$$8)9 \quad 8 \quad 15 \quad 16$$
$$\overline{3)9 \quad 1 \quad 15 \quad 2}$$
$$\overline{3 \quad 1 \quad 5 \quad 2}$$

$3 \times 5 \times 2 = 30 \times 3 \times 8 = 720$ ans.

2. What is the least number that can be divided by the nine digits without a remainder? Ans. 2520.

V. To reduce vulgar fractions to a common denominator.

RULE. Find a common denominator by the last case, in which divide each particular denominator, and multiply the quotient by its own numerator, for a new numerator, and the new numerators, being placed over the common denominator, express the fractions required in their lowest terms.

EXAMPLES.

1. Reduce $\frac{3}{4}, \frac{5}{9},$ and $\frac{7}{12}$ to a common denominator.
   36 the com. denom.

   $$\begin{array}{ll} 4 & 9 \times 3 = 27 \\ 9 & 4 \times 5 = 20 \\ 12 & 3 \times 7 = 21 \end{array}$$

   The fractions will be $\frac{27}{36}, \frac{20}{36}, \frac{21}{36}$.

2. Reduce $\frac{1}{2}, \frac{2}{3}, \frac{5}{6}$ and $\frac{7}{8}$ to a common denominator. Ans. $\frac{12}{24}, \frac{16}{24}, \frac{20}{24},$ & $\frac{21}{24}$.

3. Reduce $\frac{2}{3}, \frac{4}{9}, \frac{3}{7}$ and $\frac{5}{21}$ to a common denominator. Ans. $\frac{42}{63}, \frac{28}{63}, \frac{27}{63}$ & $\frac{15}{63}$.

4. Reduce $\frac{1}{3}, \frac{3}{5}, \frac{4}{15}$ and $\frac{5}{9}$ to a common denominator. Ans. $\frac{15}{45}, \frac{27}{45}, \frac{12}{45}$ & $\frac{25}{45}$.

VULGAR FRACTIONS.

VI. *To reduce a compound fraction to a single one.*

RULE. Multiply all the numerators for a new numerator, and all the denominators for a new denominator, then reduce the new fraction to its lowest term by Case I.

EXAMPLES.

1. Reduce $\frac{3}{4}$ of $\frac{5}{6}$ of $\frac{9}{10}$ to a single fraction.

$$\frac{3 \times 5 \times 9 = 135}{4 \times 6 \times 10 = 240} = \frac{9}{16} \text{ the answer.}$$

2. Reduce $\frac{5}{6}$ of $\frac{4}{7}$ of $1\frac{1}{2}$ to a single fraction. Ans. $\frac{55}{189}$.
3. Reduce $\frac{2}{7}$ of $\frac{5}{6}$ of $\frac{4}{5}$ to a single fraction. Ans. $\frac{8}{63}$.

VII. *To reduce a fraction of one denomination to the fraction of another, but greater, retaining the same value.*

RULE. Reduce the given fraction to a compound one, by multiplying it with all the denominations between it and that denomination, to which you would reduce it; then reduce that compound fraction to a single one.

EXAMPLES.

1. Reduce $\frac{7}{8}$ of a penny to the fraction of a pound.

$$\overset{d.}{\frac{7 \times 1 \times 1}{8 \times 12 \times 20}} = \frac{7}{1920} \text{ the answer.}$$

2. Reduce $\frac{4}{5}$ of a pennyweight to the fraction of a pound Troy. Ans. $\frac{1}{300}$.
3. Reduce $\frac{4}{7}$ of a pound Avoirdupois to the fraction of a cwt. Ans. $\frac{1}{196}$.

VIII. *To reduce the fraction of one denomination to the fraction of another, but less, retaining the same value.*

RULE. Multiply the numerator by the parts contained in the several denominations between it and that denomination to which you would reduce it for a new numerator, and place it over the denominator of the given fraction.

EXAMPLES.

1. Reduce $\frac{1}{960}$ of a pound to the fraction of a penny.

$$\frac{1 \times 20 \times 12 = 240}{960} = \frac{1}{4} \text{ the answer.}$$

VULGAR FRACTIONS.

2. Reduce $\frac{1}{300}$ of a lb. troy to the fraction of a dwt. Ans. $\frac{1}{15}$.
3. Reduce $\frac{1}{116}$ of a cwt. to the fraction of a lb. Ans. $\frac{1}{1}$.

IX. *To find the value of the fraction in the known parts of the integer.*

RULE. Multiply the numerator by the known parts of the integer and divide by the denominator.

EXAMPLES.

1. What is the value of $\frac{2}{3}$ of a £. ?

$$\frac{2}{3}\overline{)\frac{20 \text{ shillings.}}{40}}$$

Ans. 13s. 4d.

2. What is the value of $\frac{2}{3}$ of a shilling? Ans. 4d. 3⅓ qrs.
3. Reduce $\frac{3}{4}$ of a lb. troy to its proper quantity.
   Ans. 7 oz. 4 dwt.
4. Reduce $\frac{4}{5}$ of a mile to its proper quantity.
   Ans. 6 fur. 16 poles.

X. *To reduce any given quantity to the fraction of a greater denomination of the same kind.*

RULE. Reduce the given quantity to the lowest denomination mentioned for a new numerator, under which set the integral part (reduced to the same name) for a denominator, and it will express the fraction required.

EXAMPLES.

1. Reduce 16s. 8d. to the fraction of a pound.

$$\begin{array}{cc} 16 & 8 \\ 12 & \\ \hline 200 & 5 \\ \overline{240} & = \overline{6} \end{array}$$ the answer.

2. Reduce 2 quarters 3⅓ nails to the fraction of an ell English.   Ans. $\frac{5}{8}$.
3. Reduce 4s. 6½d. to the fraction of a pound.
   Ans. $\frac{109}{480}$.

ADDITION OF VULGAR FRACTIONS.

I. *To add fractions that have a common denominator.*

RULE. Add their numerators together, and place the sum over one of the given denominators.

EXAMPLES.

1. Add $\frac{1}{9}$, $\frac{2}{9}$, $\frac{4}{9}$, $\frac{5}{9}$, and $\frac{7}{9}$ together.

$$\frac{1+2+4+5+7}{9} = \frac{19}{9} = 2\frac{1}{9} \text{ the answer.}$$

2. Add $\frac{7}{24}$, $\frac{11}{24}$, and $\frac{13}{24}$ together.   Ans. $1\frac{7}{24}$.
3. Add $\frac{13}{20}$, $\frac{17}{20}$, and $\frac{9}{20}$ together.   Ans. $1\frac{19}{20}$.
4. Add $\frac{7}{16}$, $\frac{13}{16}$, and $\frac{15}{16}$ together.   Ans. $2\frac{3}{16}$.

II. *To add mixed numbers whose fractions have a common denominator.*

RULE. Add the fractions by the last case, and the integer as in whole numbers.

EXAMPLES.

1. Add $2\frac{1}{11}$, $3\frac{2}{11}$, $4\frac{4}{11}$, and $7\frac{9}{11}$ together.

$$\begin{array}{r} 2\frac{1}{11} \\ 3\frac{2}{11} \\ 4\frac{4}{11} \\ 7\frac{9}{11} \\ \hline 17\frac{5}{11} \text{ answer.} \end{array}$$

2. Add $13\frac{1}{15}$, $9\frac{4}{15}$, and $3\frac{7}{15}$ together.   Ans. $25\frac{4}{5}$.
3. Add $1\frac{1}{12}$, $2\frac{5}{12}$, $3\frac{7}{12}$, and $4\frac{11}{12}$ together.   Ans. 12.
4. Add $9\frac{13}{14}$, $7\frac{9}{14}$, $5\frac{5}{14}$, and $8\frac{11}{14}$ together.   Ans. $31\frac{4}{7}$.

III. *To add fractions, having different denominators.*

RULE. Find the least common denominator by Case III. in Reduction, in which divide each denominator, and multiply

VULGAR FRACTIONS.

the quotient by its numerator; the sum of the products is a new numerator to the common denominator, and the fraction required.

EXAMPLES.

1. Add $\frac{2}{3}, \frac{3}{4}, \frac{5}{6}, \frac{7}{8}$, and $1\frac{1}{2}$ together.

   24 com. denom.

   $$\begin{array}{r} 3 \quad 8\times 2=16 \\ 4 \quad 6\times 3=18 \\ 6 \quad 4\times 5=20 \\ 8 \quad 3\times 7=21 \\ 12 \quad 2\times 11=22 \end{array}$$

   $\frac{97}{24} = 4\frac{1}{24}$ the answer.

2. Add $\frac{1}{2}, \frac{1}{4}, \frac{1}{3}, \frac{1}{7}$, and $\frac{1}{8}$ together.   Ans. $1\frac{61}{280}$.
3. Add $\frac{4}{5}, \frac{2}{4}, \frac{6}{7}, \frac{3}{8}$, and $\frac{5}{14}$ together.   Ans. $3\frac{67}{105}$.

IV. *To add mixt numbers whose fractions have different denominators.*

RULE. Add the fractions as in the last case, and the integers as in whole numbers.

EXAMPLES.

1. Add $5\frac{2}{3}, 6\frac{7}{8}$, and $4\frac{1}{2}$ together.

   24 com. denom.

   $$\begin{array}{c|c} 5\frac{2}{3} & 16 \\ 6\frac{7}{8} & 21 \\ 4\frac{1}{2} & 12 \end{array}$$

   Ans. $17\frac{1}{24}$ $\frac{49}{24}=2\frac{1}{24}$.

2. Add $1\frac{3}{5}, \frac{2}{3}$ of $\frac{1}{3}$, and $9\frac{3}{20}$ together.   Ans. $11\frac{1}{60}$.
3. Add $1\frac{9}{10}, 6\frac{7}{8}, \frac{2}{3}$ of $\frac{1}{2}$, and $7\frac{1}{2}$ together.   Ans. $16\frac{73}{120}$.

V. *When the fractions are of several denominations.*

RULE. Reduce them to their proper quantities by Case IX. in Reduction, and add as before.

VULGAR FRACTIONS.

EXAMPLES.

1. Add $\frac{7}{9}$ of a £. to $\frac{3}{10}$ of a shilling.

$$\begin{array}{r|r} & \text{15 common measure.} \\ & s. \quad d. \\ \frac{7}{9} \text{ of a £.} = 15 \quad 6\frac{2}{3} & 10 \\ \frac{3}{10} \text{ of a } s. = 0 \quad 3\frac{3}{5} & 9 \\ \hline \text{Ans.} \quad 15 \quad 10\frac{4}{15} & \frac{12}{15} = 1\frac{4}{15} \end{array}$$

2. Add $\frac{2}{3}$ of a yard, $\frac{3}{4}$ of a foot, and $\frac{7}{8}$ of a mile together.
Ans. 1540 yds. 2 ft. 9 inches.

3. Add $\frac{1}{3}$ of a week, $\frac{1}{4}$ of a day, and $\frac{1}{2}$ of an hour together.
Ans. 2 d. 14$\frac{1}{2}$ h.

SUBTRACTION OF VULGAR FRACTIONS.

I. *To find the difference between simple fractions that have a common denominator.*

RULE. Subtract the less numerator from the greater, and under the remainder put the denominator.

EXAMPLES.

From	$\frac{5}{7}$	$\frac{11}{12}$	$\frac{15}{16}$	$\frac{17}{35}$	$\frac{105}{209}$
Take	$\frac{2}{7}$	$\frac{5}{12}$	$\frac{8}{16}$	$\frac{13}{35}$	$\frac{99}{209}$
Rem.	$\frac{3}{7}$	$\frac{1}{2}$	$\frac{7}{16}$	$\frac{4}{35}$	$\frac{6}{209}$

II. *To subtract a fraction or mixt number from a whole number.*

RULE. Subtract the numerator from the denominator, and under the remainder put the denominator, and carry one to be deducted from the integer.

EXAMPLES.

From	3	6	10	9	100
Take	$0\frac{3}{16}$	$0\frac{7}{8}$	$0\frac{1}{10}$	$5\frac{1}{2}$	$99\frac{99}{100}$
Rem.	$2\frac{13}{16}$	$5\frac{1}{8}$	$9\frac{9}{10}$	$3\frac{1}{2}$	$0\frac{1}{100}$

VULGAR FRACTIONS.

III. *To subtract simple fractions that have no common denominator.*

RULE. By Case IV, in Reduction, find a common denominator, in which divide each denominator, and multiply the quotient by its numerator; the difference between the products thus found is a numerator to the common denominator, and the answer required.

EXAMPLES.

From $\frac{17}{21}$ take $\frac{9}{14}$.

$$42 \text{ com. denom.}$$

$$21 \quad 2 \times 17 = 34$$
$$14 \quad 3 \times 9 = 27$$

Rem. $\frac{7}{42} = \frac{1}{6}$, the answer.

From	$\frac{5}{6}$	$\frac{11}{12}$	$\frac{5}{6}$	$\frac{3}{15}$	$\frac{209}{216}$
Take	$\frac{1}{2}$	$\frac{3}{4}$	$\frac{4}{5}$	$\frac{2}{20}$	$\frac{7}{144}$
Rem.	$\frac{1}{3}$	$\frac{1}{6}$	$\frac{1}{30}$	$\frac{1}{12}$	$\frac{197}{432}$

In order to distinguish the greater of two fractions, let them be reduced to a common denominator, as in case V. in Reduction; and that fraction, whose numerator is greater, is the greater fraction; the difference between the new numerators, being set over the common denominator, will shew the excess or inequality.

EXAMPLE.

Which of the two is the greater fraction, $\frac{11}{12}$ or $\frac{15}{16}$?

$$48 \text{ com. denom.}$$

$$12 \quad 4 \times 11 = 44$$
$$16 \quad 3 \times 15 = 45$$

Ans. $\frac{15}{16}$ is greater by $\frac{1}{48}$.

IV. *To subtract a fraction or mixt number from a mixt number, when the fractional part to be subtracted is greater than that from which it is to be subtracted.*

RULE. Find a common denominator and a new numerator, as in the last case, and then subtract the numerator of the greater fraction from the common denominator, and to the re-

VULGAR FRACTIONS.

mainder add the less numerator, and set the sum of both for a new numerator over the common denominator, and carry one to the integral part, and proceed as in whole numbers.

EXAMPLES.

$$\begin{array}{r|l} & 27\text{ common denominator.} \\ \text{From } 13\frac{1}{9} & 3\times 1 = 3 \\ \text{Take } 8\frac{1}{2}\frac{2}{7} & 1\times 14 = 14 \\ \hline 4\frac{1}{2}\frac{6}{7} & \frac{1}{2}\frac{6}{7} \end{array}$$

From $6\frac{5}{7}$	$10\frac{3}{10}$	$12\frac{5}{12}$	$19\frac{5}{11}$
Take $0\frac{6}{7}$	$1\frac{7}{12}$	$6\frac{1}{2}$	$0\frac{7}{15}$
Rem. $5\frac{6}{7}$	$8\frac{18}{60}$	$5\frac{11}{12}$	$18\frac{163}{165}$

V. *When the fractions are of different denominations.*

RULE. Reduce them to their proper quantities, and subtract as before.

EXAMPLES.

1. From $\frac{7}{8}$ of a £. take $\frac{3}{10}$ of a shilling.

$$\begin{array}{r|l} & 15\text{ common denominator.} \\ \frac{7}{8}\text{ of a £.}= 15s.\ 6\frac{2}{3}d. & 10 \\ \frac{3}{10}\text{ of a }s.= 0\ \ \ 3\frac{3}{5} & 9 \\ \hline \text{Rem. } 15\ \ \ 3\frac{1}{15} & \end{array}$$

2. From $\frac{3}{4}$ of a £. take $\frac{3}{4}$ of a shilling. Ans. 14s. 3d.
3. From $\frac{3}{4}$ of a lb. troy take $\frac{1}{6}$ of an ounce.
Ans. 8oz. 16dwt. 16grs.
4. From 7 weeks take $9\frac{7}{10}$ days. Ans. 5w. 4d. 7h. 12m.
5. From $\frac{1}{6}$ of a yard take $\frac{2}{3}$ of an inch. Ans. 5 inch. 1 bc.

MULTIPLICATION OF VULGAR FRACTIONS.

RULE. Reduce compound fractions to simple ones, and mixt numbers to improper fractions; then multiply the numerators together for a new numerator, and the denominators for a new denominator.

VULGAR FRACTIONS.

EXAMPLES.

1. Multiply $4\frac{1}{2}$ by $\frac{1}{8}$.

$$4\frac{1}{2}$$
$$2$$

$$\frac{9 \times 1}{2 \times 8} = \frac{9}{16}, \text{ the answer.}$$

2. Multiply $\frac{3}{8}$ by $\frac{4}{5}$.   Ans. $\frac{3}{10}$.
3. Multiply $\frac{7}{9}$ by $\frac{2}{3}$.   Ans. $\frac{14}{27}$.
4. Multiply $48\frac{3}{5}$ by $13\frac{5}{8}$.   Ans. $672\frac{3}{10}$.
5. Multiply $\frac{3}{4}$ of 9 by $\frac{7}{8}$.   Ans. $5\frac{29}{32}$.
6. Multiply $\frac{9}{10}$ by $\frac{2}{3}$ of $\frac{3}{4}$ of $\frac{5}{6}$.   Ans. $\frac{3}{8}$.

DIVISION OF VULGAR FRACTIONS.

RULE. Prepare the fractions, if necessary; then invert the divisor, and proceed as in multiplication.

EXAMPLES.

1. Divide $\frac{4}{7}$ by $\frac{2}{3}$.

$$\frac{4 \times 3}{7 \times 2} = \frac{12}{14} = \frac{6}{7} \text{ the answer.}$$

2. Divide $3\frac{1}{6}$ by $9\frac{1}{2}$.

$$\begin{array}{cc} 3\frac{1}{6} & 9\frac{1}{2} \\ 6 & 2 \end{array}$$

$$\frac{19}{6} \quad \frac{19}{2} \quad \text{Then } \frac{19 \times 2}{6 \times 19} = \frac{38}{114} = \frac{1}{3} \text{ the answer.}$$

3. Divide 5 by $\frac{7}{10}$.   Ans. $7\frac{1}{7}$.
4. Divide $\frac{9}{10}$ by $4\frac{1}{2}$.   Ans. $\frac{1}{5}$.
5. Divide $9\frac{1}{6}$ by $\frac{1}{2}$ of 7.   Ans. $2\frac{13}{21}$.
6. Divide $5205\frac{1}{2}$ by $\frac{2}{3}$ of 91.   Ans. $71\frac{1}{2}$.

MISCELLANEOUS QUESTIONS
IN THE PRECEDING RULES.

1. What part is $28\frac{11}{13}$ of $33\frac{1}{21}$?   Ans. $\frac{7}{8}$.
2. What will remain if $13\frac{5}{8}s.$ and $7\frac{3}{8}d.$ be taken from £.1?
    Ans. $5s.\ 6\frac{3}{8}d.$

VULGAR FRACTIONS.

3. Which is the greater fraction $\frac{8}{15}$ or $\frac{9}{20}$?
Ans. $\frac{8}{15}$ is greater by $\frac{1}{12}$.

4. Of what number is 176 the $\frac{11}{23}$ part? Ans. 368.

5. By how much must you multiply $13\frac{2}{3}$ that the product may be $49\frac{1}{5}$? Ans. $3\frac{3}{5}$.

6. Find two numbers so that $\frac{13}{18}$ of the one will be as much as $\frac{7}{16}$ of the other? Ans. 126 & 208, or 63 & 104, &c.

7. Which is greater, $\frac{1}{5}$ of 6s. or 1s. $2\frac{1}{2}d$.
Ans. 1s. $2\frac{1}{2}d$. is greater by $\frac{1}{10}d$.

8. A has $\frac{2}{3}$ of $\frac{3}{4}$ of a ship, and B $\frac{3}{8}$ of $\frac{4}{5}$, which is the greater share, and by how much? Ans. A's share is greater by $\frac{1}{5}$.

9. A farmer being asked, how many sheep he had, answered, that he had them in 5 fields; in the first he had $\frac{1}{4}$ of his flock, in the second $\frac{1}{6}$, in the third $\frac{1}{8}$, in the fourth $\frac{1}{12}$, and in the fifth 450; how many had he? Ans. 1200.

RULE OF THREE DIRECT IN VULGAR FRACTIONS.

RULE. Having stated the question, make the necessary preparations, as in Reduction of Fractions, and invert the first term; then proceed as in Multiplication of Fractions.

EXAMPLES.

1. If $\frac{1}{4}$ of a yard of cloth cost $\frac{2}{3}$ of a shilling, what will $\frac{7}{8}$ of a yard come to?

$$\text{If} \quad \begin{array}{c} yd. \\ \frac{1}{4} \end{array} : \begin{array}{c} s. \\ \frac{2}{3} \end{array} :: \begin{array}{c} yd. \\ \frac{7}{8} \end{array}$$

inverted

$$\frac{4 \times 2 \times 7}{1 \times 3 \times 8} = \frac{56}{24} = 2s.\ 4d.\ \text{the answer.}$$

2. If $\frac{3}{16}$ of a ship cost £273 2s. 6d. what are $\frac{5}{32}$ of her worth? Ans. £.227 12s. 1d.

3. If $\frac{1}{4}$ of a yard cost $\frac{2}{3}$ of a pound, what will $\frac{3}{5}$ of an ell English come to, at the same rate? Ans. £.2.

4. A person, having $\frac{2}{3}$ of a coal mine, sells $\frac{3}{4}$ of his share for £.171: what is the whole mine valued at? Ans. £.380.

Single Rule of Three inverse in Vulgar Fractions.

EXAMPLES.

1. If $25\frac{2}{7}s.$ will pay for the carriage of an cwt. $145\frac{1}{4}$ miles, how far may $6\frac{1}{3}$ cwt. be carried for the same money?
Ans. $22\frac{9}{26}$ miles.

2. If $3\frac{1}{4}$ yds. of cloth that is $1\frac{1}{3}$ yard wide, be sufficient to make a cloak, how much must I have of that sort, which is $\frac{4}{5}$ yard wide, to make another of the same bigness? Ans. $4\frac{7}{8}$ yds.

3. If 3 men can do a piece of work in $4\frac{1}{3}$ hours, in how many hours will 10 men do the same work? Ans. $1\frac{7}{20}$.

4. If the penny white-loaf weigh 7 oz. when a bushel of wheat cost 5s. 6d. what is the bushel worth when the penny white-loaf weighs but $2\frac{1}{2}$ oz. Ans. 15s. $4\frac{4}{5}d.$

. . . . . .

PRACTICE

Is a contraction of the Rule of Three direct, when the first term happens to be an unit, or one, and has its name from its frequent use in business.

THE TABLE.

Parts of a £.		Parts of a Ton.		Parts of $\frac{1}{2}$ Cwt.	
s. d.		Cwt. Qr.		lb.	
10	is $\frac{1}{2}$	10	is $\frac{1}{2}$	28	is $\frac{1}{2}$
6 8	...... $\frac{1}{3}$	5	.... $\frac{1}{4}$	14	...... $\frac{1}{4}$
5	...... $\frac{1}{4}$	4	.... $\frac{1}{5}$	8	...... $\frac{1}{7}$
4	...... $\frac{1}{5}$	2 2	.... $\frac{1}{8}$	7	...... $\frac{1}{8}$
3 4	...... $\frac{1}{6}$	2	.... $\frac{1}{10}$	4	...... $\frac{1}{14}$
2 6	...... $\frac{1}{8}$	1	.... $\frac{1}{20}$	$3\frac{1}{2}$	...... $\frac{1}{16}$
2	...... $\frac{1}{10}$			2	...... $\frac{1}{28}$
1 8	...... $\frac{1}{12}$	Parts of a Cwt.			
1	...... $\frac{1}{20}$	Qrs. lb.			
		2	is $\frac{1}{2}$	Parts of $\frac{1}{4}$ Cwt.	
Parts of a shilling.		1	.... $\frac{1}{4}$	lb.	
d.		16	.... $\frac{1}{7}$	14	is $\frac{1}{2}$
6	is $\frac{1}{2}$	14	.... $\frac{1}{8}$	7	...... $\frac{1}{4}$
4	...... $\frac{1}{3}$	8	.... $\frac{1}{14}$	4	...... $\frac{1}{7}$
3	...... $\frac{1}{4}$	7	.... $\frac{1}{16}$	$3\frac{1}{2}$	...... $\frac{1}{8}$
2	...... $\frac{1}{6}$	4	.... $\frac{1}{28}$	2	...... $\frac{1}{14}$
$1\frac{1}{2}$	...... $\frac{1}{8}$	2	.... $\frac{1}{56}$	1	...... $\frac{1}{28}$
1	...... $\frac{1}{12}$				

PRACTICE.

Case I.

When the price is an aliquot, or even part of a shilling.

Rule. Divide the given number by the part, and the quotient is the answer in shillings; what remains is to be reduced as in Compound Division.

Examples.

1. What will 4596 yards cost at 6d. per yard?

   6d. ½ | 4596
   ───────────
   2|0 | 229|8
   ───────────
   114 18   Ans. £.114 18s.

Yards		d.		£.	s.	d.
2. 3746	at	4	per yard. Ans.	62	8	8
3. 1095	......	3	....	13	13	9
4. 7596	......	2	....	63	6	0
5. 3747	......	1	....	15	12	3
6. 3203	......	1½	....	20	0	4½

Case II.

When the price is pence, or pence and farthings, and no even part of a shilling.

Rule. Find the even parts for the price, and proceed as in Case I. and the sum of the quotients is the answer.

Examples.

1. What will 4937 yards come to, at 9d. per yard?

   6 | ½ | 4937
   ──────────────
   3 | ½ | 2468 6
       | | 1234 3
   ──────────────
   2|0 | 370|2 9

   Ans. £.185 2 9

H 2

	Yards.		d.		£.	s.	d.
2.	2765	at	8 per yard.	Ans.	92	3	4
3.	3762		7		109	14	6
4.	3159		7½		98	14	4½
5.	1496		11		68	11	4
6.	1895		10½		82	18	1½
7.	4689½		5		97	13	11½
8.	3689		8¼		126	16	2¼
9.	1871		2½		19	9	9½
10.	8914		8¼		306	8	4½
11.	2563½		9½		101	9	5¼
12.	95¾		10½		4	3	9¼
13.	201¼		9		7	10	11¼

Case III.

When the price is shillings, or shillings and pence, and an even part of a pound.

Rule. Divide the given quantity by the even part, and the quotient is the answer in pounds. If there be a remainder, reduce it as in Compound Division.

Examples.

1. At 6*s.* 8*d.* per yard, what will 473 yards come to?

6*s.* 8*d.* | ⅓ | 473

Ans. £.157 13*s.* 4*d.*

	yards.		s.	d.		£.	s.	d.
2.	387	at	10		Ans.	193	10	0
3.	478		5			119	10	0
4.	397		3	4		66	3	4
5.	797½		2	6		99	13	9
6.	159¼		1	8		13	5	5

Case IV.

When the price is shillings or shillings and pence, which makes no even part of a pound.

Rule. Find the even parts for the price, and divide as in Case III. or multiply the given quantity by the shillings, and *take the even parts* of shillings for the pence, as in Case II.

EXAMPLES.

1. What cost 287 yards at 17*s.* 6*d.* per yard.

First method.

s.	d.		
			287
10		$\frac{1}{2}$	143 10
5		$\frac{1}{2}$	71 15
2	6	$\frac{1}{2}$	35 17 6

Ans. 251*l.* 2*s.* 6*d.*

Second method.

287
17 6
2009
287
6 | $\frac{1}{2}$ | 143 6
2|0)502|2 6

Ans. 251*l.* 2*s.* 6*d.*

	yards.		*s.*	*d.*		£.	*s.*	*d.*
2.	8172	at	15		Ans.	6129		
3.	3691	....	19		....	3506	9	
4.	4765	....	11	8	....	2779	11	8
5.	3718	....	18	4	....	3408	3	4
6.	709¼	....	12	6	....	443	5	7½
7.	213	....	14	10	....	157	19	6
8.	96½	....	2	9½	....	13	9	4¾
9.	158	....	5	8¾	....	45	5	2½
10.	4705¼	....	3	9	....	882	6	6¾
11.	127	....	7	5¼	....	47	9	10¼

CASE V.

When the price is an even number of shillings.

RULE. Multiply the quantity by half the shillings, doubling the first (or right hand) figure of the product for shillings, the rest are pounds.

EXAMPLES.

1. What will 788 yards come to, at 2 shillings per yard?

788
1 = half the shillings.

Ans. £.78 16

	yards.		*s.*		£.	*s.*
2.	347	at	4	Ans.	69	8
3.	638	....	6	....	191	8
4.	589¼	....	8	....	235	14
5.	246	....	10	....	123	0
6.	324¼	....	12	....	194	17
7.	523	....	14	....	366	2
8.	745	....	16	....	596	0
9.	373½	....	18	....	336	3
10.	270	....	20	....	270	0
11.	172¼	....	22	....	189	15
12.	89¼	....	24	....	107	2

Case VI.

When the price is pounds, shillings, &c.

Rule. Multiply the integers of the given quantity by the pounds, and work for the shillings, &c. by such of the preceding rules as you think best, and work likewise for the fractional parts of the integer; the sum of these will give the answer.

Examples.

1. What will 173 cwt. 1 qr. 14 lb. of sugar come to, at £.3 15s. 6d. per cwt. ?

```
                   173  1 14
                     3 15  6
                   ─────────
       s. d.        519
       10   | ½ |    86 10
        5   | ¼ |    43  5
            6 | 1/10 |  4  6  6

       1 qr.  | ¼ |   0 18 10½
      14 lb.  | ½ |   0  9  5¼
                   ─────────
              Ans. £.654  9  9¾
```

	cwt. qrs. lb.	s. d.	£. s. d.
2.	219 2 19	at 69 11	Ans. 767 18 6½
3.	310 3 22	.....53 8	.....834 7 5½

In working questions of this kind, when the quantity is not above the multiplication table, the following method is used.

1. What will 45 cwt. 2 qrs. 14 lb. of sugar come to, at £.3 7 9 per cwt. ?

```
                     3  7 9
                         5
                   ─────────
                    16 18 9
                         9
                   ─────────
                   152  8  9   price of 45 cwt.
       2 qrs.  ½     1 13 10½  price of 2 qrs.
      14 lb.   ¼     0  8  5¼  price of 14 lb.
                   ─────────
              Ans. £.154 11  1
```

	Tons.	*cwt.*	*qrs.*	*lb.*		*l.*	*s.*	*d.*		*l.*	*s.*	*d.*
2.		57	2	8	....	3	17	9	....	223	16	2
3.		19	3	13	....	2	5	10	....	45	10	6
4.		75	3	25	....		48	5	....	183	18	4¼
5.		2	1	18	....		59	8	....	7	3	10
6.		1	1	11	....		63	9	....	4	5	11¼
7.		0	3	19	....		54	0	....	2	9	7¾
8.	37	14	2	14	hemp	89	6	8	per ton	3370	13	2
9.	27	16	3	18	....	90	10		....	2520	0	5
10.		15	2		....	92	5		....	71	9	10½
11.	17	10	2		....	91	10		....	1603	10	9

••••••

1. What will 37 cwt. 3 qrs. 7 lb. of sugar come to, at 14 dols. 40 cts. per cwt.?

		14,40
		37
		10080
		4320
2 qrs.	½	720
1 qr.	½	360.
7 lb.	¼	90
		544,50

Ans. 544 dols. 50 cts.

	Tons.	*cwt.*	*qr.*	*lb.*		*dols.*	*cts.*			*dols.*	*cts.*	
2.	24	18	3	18	of hemp at	289	50	per ton.	Ans.	7221	73	
3.	31	16			....	268	75	....		8546	25	
4.	19	14	2	12	iron	110		....		2170	33	2
5.		17	3	24	cordage	14		per cwt.		251	50	

	A.	*R.*	*per.*		*dols.*	*cts.*			*dols.*	*cts.*
6.	25	2	25	of land at	29		per acre.	Ans.	744	3
7.	87	1	37	....	38		....		2886	88
8.	229	3	18	....	18	50	....		4252	45½
9.		3	26	....	25		....		22	81

••••••

1. How much will 49 M. 3 hund. 25 casts of staves come to, at 17 dols. per M.?

		49
		17
		343
		49
2 hund.	⅕	3,4
1	½	1,7
20 casts	½	,85
5	¼	,212
		839,162

Ans. 839 dols. 16 cts. 2 m.

94 PRACTICE.

```
   M. hun. casks.                dols.                dols. cts.
2. 19  8  15 W. O. hhd. staves   31 per M.     Ans. 614 93
3. 22  9  37 R. O.  do.  do.     13 ................ 298 90
4. 28  1   8 W. O. barrel do.    16 ................ 449 92
5.  4  2  11 ................    15 ................  63 41
           ......
```

1. What will 8,767 feet of merchantable boards come to, at 38s. 6d. per M.?

```
                8,767
                  38 6
               ───────
               70136
               26301
6d.  ½          4383
               ───────
           20)337,529 shillings.
        Ans. £.16 17 6
```

The fourth figure of the product of the remainder, multiplied by 12, is set down for pence.

```
                          s.  d.              £.  s.  d.
2. 18,370 ft. mer. boards 39  8 per M. Ans.  36   8   8
3.  2,819 do.  do.   do.  37  4 ............  5   5   2
4.   ,327 do.  do.   do.  41  0 ............  0  13   5
5.   ,183 do. refuse do.  20  6 ............  0   3   9
```

What is the amount of a seaman's wages from the 15th of March to the 6th of December following, being 8 months and 20 days, at 16 dollars per month?

```
                    16
                     8
                   ─────
                   128 for 8 months.
        15 days     8
         5         2,66⅔
                   ─────
                  138,66⅔   Ans. 138 dols. 66⅔ cts.
```

NOTE. In calculating the time of seaman's service, either of the days of engaging or being discharged is taken, but not both.

What is the amount of a seaman's wages from 15th of June to the 28th of May following, at 15 dols. per month?

Ans. 171 dols.

PRACTICE. 95

At £.4 11 3 per cwt. what will 3 qrs. 25½ lb. come to?

$$£.4 \quad 11 \quad 3$$

2 qrs.	½	2	5	7½
1 qr.	½	1	2	9¾
14 lb.	½	0	11	4⅞
7	½	0	5	8 7/16
3½	½	0	2	10 7/32
1	¼	0	0	9 87/112

Ans. £.4 9 2 125/224

What will 19 tons, 19 cwt. 3 qrs. 27½ lb. come to, at £.19 19s. 11¾d. per ton?

Ans. £.399 19s. 5 10041/17920.

....

TARE AND TRET.

TARE and TRET are allowances made in selling goods by weight.

Tare is an allowance made to the buyer for the weight of the hogshead, barrel, or bag, containing the commodity.

Tret is an allowance for waste, dust, &c. generally at 4 lb. per 104 lb.

Cloff is an allowance for the turn of the scale, at 2 lb. per 3 cwt.

Gross weight is the whole weight of the goods, together with the hogshead, barrel, or bag, &c. that contains them.

Suttle is when part of the allowance is deducted from the gross.

Neat weight is what remains after all allowances are made.

TARE AND TRET.

Custom-house allowances on tea, coffee, and sugar.

	lb.	
Tare on whole chests of bohea tea	70	Which tare shall include rope, canvass, and other coverings.
.... on every half chest do.	36	
.... on quarter do.	20	
.... on every chest of hyson, or other green teas, the gross wt. of which is 70 lb. or upwards	20	Tare for all other boxes of tea, according to invoice, or actual weight thereof.
.... on every box of other tea, not less than 50 lb. or more than 70 lb. gross	18	Tare for coffee in bags 2 per 100 .... in bales 3 do. .... in casks 12 do. On sugar, other than loaf— .... in casks 12 do.
If 80 lb. gross	20	.... in boxes 15 do.
And from 80 lb. gross and upwards	22	.... in bags or mats 5 do.

There is an allowance of two per cent. for leakage on the quantity which shall appear to be contained in any cask of liquor subject to duty by the gallon; and ten per cent. on all beer, ale, and porter in bottles, and 5 per cent. on all other liquors in bottles in lieu of breakage, or the duties may be computed on the actual quantity, at the option of the importer, to be made at the time of entry.

EXAMPLES.

1. Sold ten casks of allum, weighing gross 33 cwt. 2 qrs. 15 lb. tare 15 lb. per cask; what is the amount at 23s. 4d. per cwt.?

```
        cwt. qr. lb.
 gross  33   2   15        10 casks.
 tare    1   1   10        15 lb. per cask.
        ─────────────      ─────────────
 neat   32   1    5        112)150
                           C. 1  1  10 tare.
          Ans. £37 13 6½
```

2. At 1 dol. 25 cts. per lb. what will 3 chests of hyson tea come to, weighing gross 96 lb. 97 lb. and 101 lb.; tare 20 lb. per chest?   Ans. 292 dols. 50 cts.

TARE AND TRET.

3. At 9 dols. 49 cts. per cwt. what will 3 hhds. of tobacco come to, weighing gross, viz.

	cwt.	qrs.	lb.		lb.
No. 1.	9	3	25	tare	149
2.	10	2	12		150
3.	11	1	25		158

Ans. 265 dolls. 46¼ cents.

4. At 79s. 9d. per cwt. how much will 4 hhds. of madder come to, weighing gross, viz.

	cwt.	qrs.	lb.
No. 1.	10	3	4
2.	11	2	13
3.	10	1	16
4.	14	3	19	tare 14 lb. per cwt.

14 lb. ⅛ | 47 2 24 gross
           5 3 24 tare
          ─────────────
          41 3   0 neat.

Ans. £.166 9 6¾.

5. At 62s. per cwt. what will a hhd. of sugar come to, weighing gross 7 cwt. 1 qr.; tare 12 lb. per cwt.? Ans. £.20 1 4.

6. At 21 cents per lb. what will 6 hhds. of coffee come to, weighing gross, viz.

	cwt.	qr.	lb.		lb.
No. 1.	7	1	14	tare	96
2.	8	2	21		98
3.	7	1	21		91
4.	6	3	25		90
5.	7	0	23		89
6.	8	1	12		100

Ans. 964 dols. 32 cents.

7. What would the above coffee amount to, allowing 12 lb. per cwt. as tare on the gross weight? Ans. 966 dols. 84 cts.

8. At 72s. 6d. per cwt. how much will 8 hhds. of sugar come to, weighing gross each 8 cwt. 3 qrs. 7 lb.; tare 12 lb. per cwt.?
Ans. £.228 3 7¼.

9. At 23 cents per lb. what will 4 bags of coffee come to, weighing gross 450 lb.; tare 2 per cent. or 2 lb. per 100 lb.?
Ans. 101 dols. 43 cents.

10. At 12 dols. 50 cents per cwt. what will 3 barrels of sugar come to, weighing gross, viz.

	cwt.	qrs.	lb.
No. 1.	2	2	10
2.	2	1	21
3.	2	0	15	Tare 21 lb. per barrel.

Ans. 82 dols. 47 cts. 7 m.

I

TARE AND TRET.

11. At 15 dols. 40 cts. per cwt. what will 4 hhds. of sugar come to, weighing gross, viz.

	cwt.	qrs.	lb.
No. 1.	7	3	13
2.	8	1	10
3.	7	2	12
4.	8	1	21

Tare 12 lb. per cwt.
Ans. 443 dols. 43 cts. 7 ms.

12. A has in his possession a hhd. of sugar, weighing gross 9 cwt. 3 qrs. owned equally between him and B. It is required to know how much sugar he should weigh out to B, allowing tare 12 lb. per cwt. ? Ans. 4 cwt. 1 qr. 11½ lb.

13. At 19½ cents per lb. what will 2 hhds. of coffee come to, weighing gross 15 cwt. 3 qrs. 11 lb. allowing custom-house tare or 12 lb. per 100 ?

```
                  15   3   11
                  ─────────────
                  1500 = fifteen hundred.
                   180 = 15×12 for excess in each cwt.
                    84 = three quarters.
                    11
                  ─────
        Gross     1775                  1775
        Tare       213         Tare       12 per 100.
                  ─────                 ──────
        Neat      1562                  213,00
                  19¼
                  ─────
                 14058
                  1562
                   781
                 ──────
                 30459 cts.        Ans. 304 dols. 59 cts.
```

14. B buys of C a hogshead of Coffee, weighing gross 9 cwt. 2 qrs. tare 12 lb. per cwt. what will it amount to at 23 cents per lb. ? Ans. 218 dols. 50 cents.

15. If custom-house tare, or 12 lb. per 100, were allowed on the above coffee, what would it amount to, and what difference would it make to the buyer ?
Ans. It amounts to 215 dols. 51 cts. being 2 dols. 99 cts. in his favour.

16. What is the gross weight of a hogshead of tobacco, weighing neat 11 cwt. 1 qr. tare 14 lb. per cwt. ?
Ans. 12 cwt. 3 qrs. 12 lb.

SINGLE FELLOWSHIP. 99

FELLOWSHIP

Is when two or more join their stocks and trade together, dividing their gain or loss, in proportion to each person's share in the joint stock.

SINGLE FELLOWSHIP.

Single Fellowship is when different stocks are employed for a certain equal time.

RULE. As the whole stock is to the whole gain or loss, so is each man's particular stock to his particular share of the gain or loss.

EXAMPLES.

1. A and B buy certain merchandizes, amounting to £.120, of which A pays £.80 and B £.40, and they gain by them £.32—what part of it belongs to each?

A £.80
B 40

As 120 : 32 :: { 80   Ans. £.21 6 8 A.
                 40   10 13 4 B.

2. A ship worth 8400 dollars being lost at sea, of which $\frac{1}{4}$ belonged to A, $\frac{1}{3}$ to B, and the remainder to C, what loss will each sustain, supposing they have 6000 dollars insured?

Ans. A's loss 600, B's 800, and C's 1000 dols.

3. A and B have gained 1260 dollars, whereof A is to have 10 per cent. more than B, what is the share of each?

Ans. A's 660, B's 600 dols.

4. A bankrupt is indebted to A 500 dols. 37 cts. to B 228 dols. to C 1291 dols. 23 cts. to D 709 dols. 40 cts. and his estate is worth but 2046 dols. 75 cts. how much does he pay per cent. and what is each creditor to receive?

Ans. He pays 75 per cent. and A's part is 375 dols. 27$\frac{3}{4}$ cts. B's 171 dols. C's 968 dols. 42$\frac{1}{4}$ cts. and D's 532 dols. 5 cts.

5. Three boys, John, James, and William, buy a lottery ticket for 3 dols. of which John pays 90 cts. James 1 dol. and William the remainder. This ticket is entitled to a prize of 2000 dollars, subject to a deduction of 12$\frac{1}{2}$ per cent. how much is each to receive?

Ans. John 525 dols. James 583 dols. 33$\frac{1}{3}$ cts. William 641 dols. 66$\frac{2}{3}$ cts.

6. Three merchants made a joint stock—A put in £.565 6 8, B £.478 5 4, and C a certain sum, and they gained £.373 9 11, of which C took for his part £.112 11 11—required A and B's part of the gain, and how much C put in?

Ans. A's gain £.141 6 8, B's £.119 11 4, and C put in £.450 7 8.

7. Three men have to share a legacy of 1500 dols. of which B is to have ½, C ¼ and D the remainder, but C relinquishes his part to B and D, leaving it to be divided between them, according to their shares in the whole. It is required to know how much of the legacy B and D receive respectively?

Ans. B's part is 1000, D's 500 dols.

. . . .

DOUBLE FELLOWSHIP.

Double Fellowship is when the stocks are employed for different times.

RULE. Multiply each man's stock by its time, and add them together, then say, As the sum of the products is to the whole gain or loss, so is the product of each man's stock and time to his share of the gain or loss.

EXAMPLES.

1. B and C trade in company, B put in £.950 for 5 months, and C £.785 for 6 months, and by trading they gain £.275 18 4; what is each man's part of the profit?

B's stock 950 × 5 = 4750
C's   785 × 6 = 4710
            ———
As 9460 : 275 18 4 :: { 4750 : l.138 10 10 B's.
                      { 4710 : 137 7 6 C's.

2. Two merchants enter into partnership for 16 months. A put into stock at first 1200 dols. and at the end of 9 months 200 dols. more, B put in at first 1500 dols. and at the expiration of 6 months took out 500 dols.—with this stock they gained 772 dols. 20 cts. what is each man's part of it?

Ans. A's 401 dols. 70 cts.—B's 370 dols. 50 cts.

3. Two persons hired a coach in Boston, to go 40 miles, for 20 dols. with liberty to take in 2 more when they pleased. Now when they had gone 15 miles, they admit C, who wished to go the same route; and on their return, within 25 miles of Boston, they admit D for the remainder of the journey. Now as each person is to pay in proportion to the distance he rode, it is required to settle the coach-hire between them.

Ans. A and B 6 dols. 40 cts. each, C 5 dols. 20 cts. and D 2 dols.

SIMPLE INTEREST.

Is a compensation made by the borrower of any sum of money to the lender, according to a certain rate per cent. agreed on for the principal only.

The legal rate of interest in Massachusetts is 6 per cent.

Principal, is the money lent.
Rate, is the sum per cent. agreed on.
Amount, is the principal and interest added together.

GENERAL RULE. Multiply the principal by the rate per cent. and divide the product by 100, and the quotient is the answer for one year.

EXAMPLES.

1. What is the interest of £.496 for one year at 6 per cent.?

```
      496
        6
     ----
    29|76
       20
     ----
    15|20
       12
     ----
     2|40
        4
     ----
     1|60         Ans. 29l. 15s. 2¾d.
```

2. What is the interest of £.383 15 9 for 2 years, at 8½ per cent.?

```
    383 15  9
          8½
    ---------
   3070  6  0
    191 17 10½
    ---------
  32|62  3 10½
    20
    ---------
  12|43
    12
    ---------
   5|26
     4
   ---------
   1|06         32l. 12s. 5¼d. for one year.
                           2
                ---------------------
                Ans. 65 4 10½ for 2 years.
```

I2

SIMPLE INTEREST.

3. What will £.826 13 9 amount to in 1 year at 5 per cent.?
$5 = \frac{1}{20}$)826 13 9 principal.
        41 6 8¼ interest.

Ans. £.868 0 5¼ amount.

4. What is the interest of £.103 11 4, for 4 years, at 7½ per cent. per annum ? Ans. £.31 1 4½.

5. What will £.36 14 9 amount to, in 3 years, at 5 per cent. per annum ? Ans. £.42 4 11¼.

6. What is the amount of £.19 15 8, for 5 years, at 6¾ per cent. per annum ? Ans. £.26 9 1¼.

7. How much is the interest of £.72 12 6, for 6 months, at 6 per cent. per annum.

    72 12 6
         6

4|35 15 0
   20

7|15
  12

1|80   l. s. d.
   4   6 m. ½)4 7 1¼ for 1 year.

3|20   Ans. 2 3 6¼ for 6 months.

NOTE. When the time is months, multiplying by the rate for the time gives the answer. This rate is found by multiplying the time by the given rate per cent. for a year, and dividing the product by 12. The quotient is the rate required, and is always equal to half the months when the yearly rate is 6 per cent.

8. What is the interest of £.25 19 3 for 8 months, at 6 per cent. per annum ?

8 months.   25 19 3
  6   4

12)48   1,03 17 0
                           20
4 rate = half the months.
                           0,77
                           12

                           9,24

Ans. £.1 0 9.

SIMPLE INTEREST. 103

9. How much will £.53 9 4 amount to, in 20 months, at 6 per cent. per annum ? Ans. £.58 16 3.

10. How much is the interest on a bond of £.295 17 10, for 18 months, at 8 per cent. per annum ?

```
    18              295  17  10
     8                       12  the rate for the time.
             
12)144             35,50  14   0
    12                    20
                       ─────────
                       10,14
                          12
                       ─────────
                        1,68
                           4
                       ─────────
                        2,72            Ans. 35l. 10s. 1¼d.
```

11. How much is the interest of £.80 12 9, for 23 months, at 6 per cent. per annum ? Ans. £.9 5 5¼.

12. How much is the interest of £.36 14 9 from 19th May to 25th October, at 6 per cent. ?

```
   36  14  9          4 m. = ⅓)2  4  1 for 1 year.
        6                        ─────────
   ─────────                     0 14  8¼
   2,20   8  6         1 m. = ¼  0  3  8
     20                6 d. = ½  0  0  8¼
   ─────────                  
   4,08                         Ans. 0 19  1
     12
   ─────────
   1,02
```

13. What will £.187 14 9 amount to, from 11th June, 1797, to 26th October, 1798, at 6 per cent. per annum ? Ans. £.203 4 5¾.

14. How much is the interest of £.19 13 7 from 3d January, 1797, to 18th May, 1798, at 6 per cent. per annum ? Ans. £.1 12 5¼.

To find the interest of any sum for months, at 6 per cent. per annum, by contraction.

RULE. Multiply the pounds by the number of months; the first or units figure of the product is pence, and the rest are shillings, observing to increase the pence in the product by 1 when they exceed 4.

SIMPLE INTEREST.

EXAMPLES.

15. What is the interest of £.56 for 1, 5, 7, and 12 months?

	56	56	56	56
mo.	1	5	7	12
Ans.	5s. 7d.	28s. 0d.	39s. 2d.	67s. 2d.

16.	£. 45	for	6 months.	Ans. £.1	7	0
17.	324		5	8	2	0
18.	19		7	0	13	3
19.	11		1	0	1	1¼

If there are Shillings, &c.

To the pounds add the decimal of the nearest even number of shillings (this will be sufficiently exact for business) and multiply by the months as before, separate two figures of the product to the right, and the left hand figures are the shillings, then multiply the figures pointed off, by 12, and the product, rejecting two figures to the right, is the pence of the answer.

| 2 | 4 | 6 | 8 | 10 | 12 | 14 | 16 | 18 shillings. |
| ,1 | ,2 | ,3 | ,4 | ,5 | ,6 | ,7 | ,8 | ,9 decimals. |

20. How much is the interest of £.347 5 9 for 3 months?

```
         347,3
             3
shillings 104,19
         ───────
    Ans. 5l. 4s. 2d.
```

21. How much is the interest of £.195 15 10½ for 10 months?

```
          195,8                ,80
            10                  12
shillings 195,80              ─────
                              9,60
    Ans. 9l. 15s. 9½d.           4
                              ─────
                              2,40
```

The value of the remainder is thus shewn to be 9½d.

SIMPLE INTEREST. 10

22. What is the interest of £.590 19 9¾ for 3 years, months and 19 days?

$$£.591 \text{ nearly.}$$
$$43$$
$$\overline{1773}$$
$$2364$$
15 days ½ 295
3   ⅕ 59
1   ⅓ 19
$$\overline{2578,6 + 1 \text{ because it exceeds 4--see the Rule}}$$
$$£.128 \; 18 \; 7$$

23. How much is the interest of £.476 9 8 for 9 month and 13 days?

$$476,5$$
$$9$$
$$\overline{4288,5}$$
10 days ⅓ 158,8
3 do.   1/10 47,6
$$\overline{449,49}$$

Ans. £.22 9 5¾

24. What is the interest of £.40 for 7 years, 5 months, and 26 days?

$$40$$
$$89 \text{ months.}$$
$$\overline{3560}$$
15 days ½ 20
10 do. ⅓ 13
1 do.   1/10 1
$$\overline{359,4}$$

Ans. £.17 19 5

SIMPLE INTEREST.

25. What is the interest of £.240 for 50 days, at 6 per cent.?

Or by Compound Proportion.

```
      240                    240
        6                     50
      ———                   ———
    14,40              6083)12000(1
      20                    6083
      ———                   ———
     8,00                    5917
                              20
   d.       d.              ———
365 : 14l. 8s. :: 50 : 1l. 19s. 5¼d.   6083)118340(19
                              6083
                              ———
                             57510
                             54747
                              ———
                              2763
                                12
                              ———
                         6083)33156(5
                              30415
                              ———
                              2741
                                 4
                              ———
                         6083)10964(1
                               6083
                               ———
                               4881
```

Ans. £.1 19 5¼.

SIMPLE INTEREST IN FEDERAL MONEY.

The principal given in English money, and the interest required in federal.

RULE. Reduce the given sum to shillings, the product gives the answer in cents, and the pence are mills nearly; the reason is, that at 6 per cent. per annum, one fifth of a dollar is the annual interest of a pound; that is, 20 cents for 20 shillings, or 1 cent for every shilling in any given sum.

EXAMPLES.

1. Required the interest of £.194 15 3 for 1 year in federal money.

```
      194 15 3
       20
      ———
     3895 cents.        Ans. 36 dols. 95 cts. 3 mills.
```

SIMPLE INTEREST. 107

2. What is the interest of £.129 13 2 for 2 years in federal money?

   129 13 2
    20
   ―――――
   2593,2 for 1 year.
     2
   ―――――
   5186,4   Ans. 51 dols. 86 cts. 4 ms.

3. What is the interest of £.91 12 1 for 5 years, in federal money?

   91 12 1
    20
   ―――――
   1832,1 for 1 year.
     5
   ―――――
   91,605 for 5 years.   Ans. 91 dols. 60½ cts.

4. What is the interest of £.139 17 2 for 4 months?

   139 17 2
    20
  ―――――――
4 mo. ⅓ | 2797,2
  ―――――――
   9,32,4   Ans. 9 dols. 32 cts. 4 ms.

* * * * * *

Principal in federal money, and Interest required in the same.

RULE. Multiply the principal by the rate per cent. and as you suppose 100 for a divisor, point off the quotient as in division of decimals.

The following rule answers the same purpose.

When the principal is dollars only, multiply by the rate, and from the product point off two figures to the right, the figures to the left hand of the point give the answer in dollars, and the rest are decimal parts or cents.

If there are cents, &c. in the principal, multiply by the rate and point off as above. The figures to the left of the point give the answer in the same name with the lowest denomination in the principal.

SIMPLE INTEREST.

EXAMPLES.

5. What is the interest of 419 dollars for 1 year at 6 per cent?

```
  419
    6
  ———
25,14          Ans. 25 dols. 14 cts.
```

6. What is the interest of 173 dollars 50 cents for 1 year, at 6 per cent?

```
        173,50
             6
        ———————
Cents 1041,00   Ans. 10 dols. 41 cts.
```

7. What is the interest of 327 dols. 82 cts. 5 mills, for 1 year, at 8 per cent?

```
        327,82,5
               8
        ——————————
mills  26226,00
                Ans. 26 dols. 22 cts. 6 ms.
```

8. How much is the interest of 325 dollars for 3 years, at 6 per cent. per annum?

```
    325                  Or thus,   325
      6                             18  rate for the time.
   ————                             ————
  19,50 for 1 year.                 2600
      3                              325
   ————                             ————
  58,50 for 3 years.                58,50
                                    Ans. 58 dols. 50 cts.
```

* * * * * *

When the time is months.

RULE. Multiply by half the number; this, as was before observed, is always equal to the rate, for the time, when the annual rate is 6 per cent. per annum.

EXAMPLES.

9. What is the interest of 284 dollars, for 8 months, at 6 per cent.?

```
       284
         4
       ————
      11,36      Ans. 11 dols. 36 cts.
```

SIMPLE INTEREST.

10. How much is the interest of 187 dols. 25 cts. for 16 months, at 6 per cent. per annum?

$$\begin{array}{r} 187,25 \\ 8 \\ \hline \end{array}$$

Cents 1498,00   Ans. 14 dols. 98 cts.

11. What is the interest of 95 dollars, for 2 months, at 6 per cent. per annum?

$$\begin{array}{r} 95 \\ 1 \\ \hline ,95 \end{array}$$

Ans. 95 cents.

12. How much is the interest of 126 dollars, 46 cents, for 9 months, at 6 per cent.?

$$\begin{array}{r} 126,46 \\ 4\tfrac{1}{2} \\ \hline 505,84 \\ 63,23 \\ \hline \end{array}$$

Cents 569,07   Ans. 5 dols. 69 cts.

13. How much is the interest of 124 dollars, for 1 month, at 6 per cent.?

$\tfrac{1}{2}$)124   Or 124
 ,62   ,5
                       ,62,0   Ans. 62 cts.

14. What is the interest of 694 dols. 84 cts. for 9 months, at 10 per cent. per annum?

$$\begin{array}{r} 694,84 \\ 10 \\ \hline \end{array}$$

Cents 6948,40 for a year.   Or 694,84
                                     7$\tfrac{1}{2}$ = rate for the time.
6 $\tfrac{1}{2}$ 3474,2   4863,88
3 $\tfrac{1}{4}$ 1737,1   347,42
                                    Cents 52,11,30
      52,11,3
                                  Ans. 52 dols. 11 cts. 3 m.

K

SIMPLE INTEREST.

15. How much is the amount of 985 dollars, for 5 years and 8 months, at 6 per cent. per annum?

```
       dols.
        985
         34 half the months.
        ----
       3940
       2955
       -----
       334,90 interest.
       985,   principal.
       --------
      1319,90 amount.      ▲ Ans. 1319 dols. 90 cts.
```

• • • • • •

When the time is months and days, and the annual rate 6 per cent. Multiply by half the months and one sixth of the days, which is equal to the rate, for the given time, and separate one figure to the right for the decimal in the rate, and proceed as usual. Should there be a remainder in taking a sixth of the days, reduce it to a vulgar fraction; this, and not the decimal, will *always* give the exact rate.

EXAMPLES.

16. What is the interest of 194 dols. for 4 months and 12 days, at 6 per cent.?

```
                    dols.
                    194
  m.     m.         2,2 =to the rate, found by the rule,
 12 : 6 :: 4,4      ----   or the annexed calculation.
           6        388
          ----      388
       12)26,4      ------
          ----      4,26,8
           2,2                 Ans. 4 dols. 26 cts. 8 ms.
```

17. How much is the interest of 263 dollars, 48 cents, for 2 months and 21 days, at 6 per cent.?

```
        dols. cts.
        263,48
          1,3½
        -------
         79044
         26348
         13174
        -------
   Cents 355,698          Ans. 3 dols. 55 cts. 6 ms.
```

SIMPLE INTEREST.

18. How much is the interest of 318 dols. for 10 months and 16 days, at 6 per cent.?

$$\begin{array}{r} 318 \\ 5,2\tfrac{2}{3} \\ \hline 636 \\ 1590 \\ \tfrac{3}{6}\quad 106 \\ \tfrac{1}{3}\quad 106 \\ \hline \text{dols.}\;16,74,8 \end{array}$$

Ans. 16 dols. 74 cts. 8 m.

19. What is the interest of 418 dols. for 1 year, 7 months and 17 days, at 6 per cent.?

$$\begin{array}{r} 418 \\ 9,7\tfrac{3}{6} \\ \hline 2926 \\ 3762 \\ \tfrac{3}{6}\text{ or }\tfrac{1}{2}\quad 209 \\ \tfrac{2}{6}\text{ or }\tfrac{1}{3}\quad 139 \end{array}\Big\} = 348\tfrac{1}{3}$$

$$\begin{array}{r} 418 \\ 5 \\ \hline 6)2090 \\ \hline 348\tfrac{1}{3} \end{array}$$

dols. 40,89,4   Ans. 40 dols. 89 cts. 4 m.

20. How much is the interest of 268 dols. 44 cts. for 3 years, 5 months, and 26 days, at 6 per cent.?

$$\begin{array}{r} 268,44 \\ 20,9\tfrac{2}{3} \\ \hline 241596 \\ 536880 \\ \tfrac{1}{3}\quad 8948 \\ \hline \text{Cents}\;5619,34,4 \end{array}$$

Ans. 56 dols. 19 cts. 3 m.

21. What is the interest of 1 dollar, for 18 days, at 6 per cent.?

$$\begin{array}{r} 1 \\ ,3 \\ \hline ,00,3\;\text{mills.} \end{array}$$

Ans. 3 mills.

· One figure is separated for the decimal in the multiplier, and two cyphers are supplied and pointed, according to the general rule.

SIMPLE INTEREST.

22. What is the interest of 910 dols. 50 cts. for 3 years, 9 months, and 26 days, at 7 per cent. per annum?

```
            910,50              Or thus, 910,50
               7                          22,9⅓
            ──────                        ──────
            63,73,50                   819450
               3                       182100
            ──────                     182100
            191,20,5 for 3 years.       30350
6 mo.  ½    31,86,7                   ──────
3 mo.  ½    15,93,3              ⅙)208,80,80|0 at 6 per cent.
15 days ¼    2,65,5                   34,80,1
10 days ⅙    1,77,0                   ──────
1 day  1/10   ,17,7              dols. 243,60,9 at 7 per cent.
            ──────
   dols.   243,60,7              Ans. 243 dols. 60 cts. 8 ms.
```

23. How much will 185 dols. 26 cts. amount to, in 2 years, 3 months, and 11 days, at 7½ per cent. per annum?
Ans. 216 dols. 94 cts. 4 ms.

24. What is the interest of 57 dols. 78 cts. for 1 year, 4 months, and 17 days, at 4 per cent. per annum?
Ans. 3 dols. 19 cts.

25. How much is the amount of 298 dols. 59 cts. from 19th May, 1797, to the 11th of August, 1798, at 8 per cent. per annum?
Ans. 327 dols. 98 cts. 4 ms.

26. How much is the amount of 196 dollars, from June 14, 1798, to April 29, 1799, at 5¾ per cent. per annum?
Ans. 205 dols. 86 cts.

27. What is the interest of 658 dollars, from January 9 to October 9 following, at ½ per cent. per month?
Ans. 29 dols. 61 cts.

In the calculation of interest in federal money, thus far, the year is supposed to be 12 months of 30 days each, making it only 360 days. Most persons use this method of computing the time, but as it is 5 days less in a year than the true number, some merchants calculate by days, without any regard to months, as being more accurate.

SIMPLE INTEREST.

EXAMPLES.

28. What is the interest of 7086 dollars, for 39 days, at 6 per cent. per annum?

By Compound Proportion.
```
      7086
        39
      ————
     63774
     21258
            ——dols cts.
    6083)276354(45-43
         24332
         ————
          33034
          30415
          ————
          26190
          24332
          ————
           18580
           18249
           ————
             331        Ans. 45 dols. 43 cts.
```

29. What is the interest of 87 dols. 56 cts. for 72 days, at 6 per cent. per annum?

```
        87,56
           72
        ————
        17512
        61292
               cts. m.
    6083)6304,32(103 6
         6083
         ————
          22132
          18249
          ————
           38830
           36498
           ————
            2342   Ans. 1 dol. 3 cts. 6 m.
```

	dols. cts.	days.		dols. cts. m.
30.	2962 19	for 254 at 6 per cent. per ann. Ans.	123 68 8
31.	35	256	1 47 2
32.	1733 97	102	29 7 5
33.	435 52	47	3 51 9
34.	215 80	125	4 43 4
35.	517 90	84	7 15 1
36.	73 63	92	1 11 3

K 2

SIMPLE INTEREST.

The following method of calculating the interest upon accounts, when there are partial payments, is sometimes used.

```
1798.                    dols.           days.    Prod. princ. & time.
January  2, Lent——100 on interest for 13 ············ 1300
————— 15, Lent——110
                         ———
                         210 ··············,  5 ·········· 1050
————— 20, Received 162
                         ———
                          48 ··············14 ········· 672
February 3, Lent—— 95
                         ———
                         143 ············· 7 ········· 1001
————— 10, Received 90
                         ———
                          53 ············· 6 ·········· 318
————— 16, Lent——186
                         ———
                         239 ·············10 ········· 2390
————— 26, Received 70
                         ———
                         169 ············· 3 ·········· 507
March    1, Lent——250
                         ———
                         419 ············· 2 ·········· 838
————— 3, Received 270
                         ———
                         149 ·············10 ········· 1490
————— 13, Received 143
                         ———
        20, Time of adjustment  6 ··········· 7 ········  42
                                                         ————
                                                         9608
```

d.cts.
Then 6083)9608(1,57 interest at 6 per cent.
        6083   6, the principal due.
        ———— ————
        35250 7,57 the amount due March 20th.
        30415
        ————
        48350
        42581
        ————
        5769

SIMPLE INTEREST. 115

By this method interest may be calculated on accounts, multiplying each sum by the days it is at interest, and taking the quotient of 36500, divided by the rate per cent. as a fixed divisor to the sum of the products. Thus, the rate in the last example being 6 per cent. the divisor is 6083 ; for 5 per cent. it would be 7300 ; for 7 per cent. 5214, &c.

If the time is *months*, multiply each sum by the months it is at interest, and take the quotient of 1200, divided by the rate as a divisor. Thus, for 6 per cent. the divisor is 200; for 5 per cent. 240; for 8 per cent. 150, &c.—(*See Compound Proportion.*)

. . . .

IN COMPUTING INTEREST ON NOTES, &c.

It has generally been the custom to find the amount of the principal from the time the interest commenced to the time of settlement, and likewise the amount of each payment, and then deduct the amount of the several payments from the amount of the principal.

EXAMPLE.

A, by his note dated April 25th, 1798, promises to pay to B 774 dols. 76 cts. on demand, with interest to commence 4 months after the date. On this note are the following endorsements:

Received, *Oct.* 12th, 1798, 260 dols. 19 cts.—*Oct.* 13th, 1798, 60 dols.—*Nov.* 2, 1798, 200 dols. And the settlement is made *Dec.* 15th, 1798.

CALCULATION.

	dols. cts.
The principal carrying interest from 25th Aug. 1798	774 76
Interest to Dec. 15, 1798 · · · · · 3 m. 20 days	14 20
Amount of the principal	788 96

	dols. cts.
First payment, Oct. 12th, 1798	260 19
Interest to Dec. 15th, 1798 · · · · 2 ms. 3 days	2 73
Second payment, Oct. 13th, 1798	60 00
Interest to Dec. 15th, 1798 · · · · 2 ms. 2 days	0 62
Third payment, Nov. 2, 1798	200 00
Interest to Dec. 15, 1798 · · · · 1 m. 13 days	1 43
Amount of the payments	524 97
Settlement is made for	Dollars—263 99

SIMPLE INTEREST.

RULE established by the Courts of Law in Massachusetts for making up judgments on SECURITIES FOR MONEY, *which are upon Interest, and on which partial payments have been endorsed.*

COMPUTE the interest on the principal sum, from the time when the interest commenced to the first time when a payment was made, which exceeds either alone or in conjunction with the preceding payments (if any) the interest at that time due: add that interest to the principal, and from the sum subtract the payment made at that time, together with the preceding payments (if any) and the remainder forms a new principal; on which compute and subtract the interest, as upon the first principal: and proceed in this manner to the time of the judgment. By this Rule, the payments are first applied to keep down the interest; and no part of the interest ever forms a part of a principal carrying interest.

The following example will illustrate the rule, in which the interest is computed at the rate of 6 per cent. by the year, that being the legal rate of interest in Massachusetts.

A, by his note dated January 1, 1780, promises to pay B 1000 dols. in six months from the date, with interest from the date.

On this note are the following endorsements:

Received, *April* 1, 1780, 24 dols.—*August* 1, 1780, 4 dols.—*Dec.* 1, 1780, 6 dols.—*Feb.* 1, 1781, 60 dols.—*July* 1, 1781, 40 dols.—*June* 1, 1784; 300 dols.—*Sept.* 1, 1784, 12 dols.—*Jan.* 1, 1785, 15 dols. and *Oct.* 1, 1785, 50 dols.—and the judgment is to be entered *Dec.* 1, 1790.

CALCULATION.

	dols. cts.
The principal sum carrying interest from January 1, 1780	1000 00
Interest to April 1, 1780, 3 months.	15 00
Amount,	1015 00
Paid April 1, 1780, a sum exceeding the interest	24 00
Remainder for a new principal	991 00
Interest on 991 dols. from April 1, 1780, to Feb. 1, 1781, (10 mo.)	49 55
Amount	1040 55
Paid Aug. 1, 1780, a sum less than the interest then due Dls. 4 00	
Paid Dec. 1, 1780, do. do. 6 00	
Paid Feb. 1, 1781, do. greater than the interest then due 60 00	
	70 00

SIMPLE INTEREST.

	dols. cts.
Remainder for a new principal	970 55
Interest on 970 dols. 55 cts. from Feb. 1, 1781, to July 1, 1781, (5 months)	24 26
Amount	994 81
Paid July 1, 1781, a sum exceeding the interest	40 00
Remainder for a new principal	954 81
Interest on 954 dols. 81 cts. from July 1, 1781, to June 1, 1784, (2 years 11 months)	167 09
Amount	1121 90
Paid June 1, 1784, a sum exceeding the interest	300 00
Remainder for a new principal	821 90
Interest on 821 dols. 90 cts. from June 1, 1784, to Oct. 1, 1785, (1 year 4 months)	65 75
Amount	887 65
Paid Sept. 1, 1784, a sum less than the interest then due, Dls.	12 00
Paid Jan. 1, 1785, do. do.	15 00
Paid Oct. 1, 1785, do. greater with two last payments than interest then due	50 00
	77 00
Remainder for a new principal	810 65
Interest on 810 dols. 65 cts. from Oct. 1, 1785, to Dec. 1, 1790, the time when judgment is to be entered (5 years 2 months)	251 30
Judgment rendered for the Amount	1061 95

······

A TABLE,

Shewing the number of Days, from any Day in any Month, to the same Day in any other Month, through the Year.

From	Jan.	Feb.	Mar.	Ap.	May.	Jun.	July.	Aug.	Sep.	Oc.	Nov.	Dec.
To Jan.	365	334	306	275	245	214	184	153	122	92	61	31
Feb.	31	365	337	306	276	245	215	184	153	123	92	62
Mar.	59	28	365	334	304	273	243	212	181	151	120	90
April	90	59	31	365	335	304	274	243	212	182	151	121
May	120	89	61	30	365	335	304	273	242	212	181	151
June	151	120	92	61	31	365	335	304	273	243	212	182
July	181	150	122	91	61	30	365	334	303	273	242	212
Aug.	212	181	153	122	92	61	31	365	334	304	273	243
Sept.	243	212	184	153	123	92	62	31	365	335	304	274
Oct.	273	242	214	183	153	122	92	61	30	365	334	304
Nov.	304	273	245	214	184	153	123	92	61	31	365	335
Dec.	334	303	275	244	214	183	153	122	91	61	30	365

SIMPLE INTEREST.

THE USE OF THE TABLE.

Suppose the number of days between the 3d of May and the 3d of November was required; look in the column under May for November, and against that month you will find 184.

If the given days be different, it is only adding or subtracting their inequality to or from the tabular number. Thus, from May 3d to Nov. 17th is $184+14=198$ days, and from Nov. 17th to May 3d is $181-14=167$ days.

If the time exceed a year, 365 days must be added; thus from the 4th of February, 1798, to the 4th of Sept. 1799, is $212+365=577$ days.

NOTE. In leap-years, if the end of the month of February be in the time, one day must be added on that account.

* * * * * *

COMPOUND INTEREST

Is that which arises both from the *principal* and *interest*; that is, when the interest on money becomes due, and not paid; it is added to the principal, and interest is calculated on this amount as on the principal before.

RULE. Find the simple interest of the given sum for one year, and add it to the principal, and then find the interest for that amount for the next year, and so on for the number of years required. Subtract the principal from the last *amount*, and the remainder will be the compound interest.

EXAMPLES.

1. What is the interest of £.246 14s. 6d. for 3 years, at 6 per cent. per annum?

5	$\frac{1}{20}$	246	14	6	
1	$\frac{1}{3}$	12	6	8½	} first year's interest.
		2	9	4	
5	$\frac{1}{20}$	261	10	6¼	amount of the first year.
1	$\frac{1}{3}$	13	1	6¼	} second year's interest.
		2	12	3½	
5	$\frac{1}{20}$	277	4	4¼	amount of the second year.
1	$\frac{1}{3}$	13	17	2½	} third year's interest.
		2	15	5¼	
		293	17	0	amount of the third year.
		246	14	6	first principal.
		47	2	6	compound interest for 3 years.

Ans. £.47. 2s. 6d.

COMPOUND INTEREST.

2. What is the compound interest of £.760 10s. for 4 years, at 6 per cent. per annum ?   Ans. £.199 12s. 2d.

3. How much is the amount of £.128 17s. 6d. for 6 years, at 6 per cent. per annum, compound interest ?
                                  Ans. £.182 16 2¾.

4. How much is the amount of 500 dollars, for 3 years, at 6 per cent. per annum, compound interest ?

$$
\begin{array}{c|c}
5 \quad \tfrac{1}{20} & 500, \\
1 \quad \tfrac{1}{5} & \left. \begin{array}{c} 25, \\ 5, \end{array} \right\} \text{first interest.} \\
\\
5 \quad \tfrac{1}{20} & 530, \\
1 \quad \tfrac{1}{5} & \left. \begin{array}{c} 26,50 \\ 5,30 \end{array} \right\} \text{second interest.} \\
\\
5 \quad \tfrac{1}{20} & 561,80 \\
1 \quad \tfrac{1}{5} & \left. \begin{array}{c} 28,09 \\ 5,61\tfrac{3}{4} \end{array} \right\} \text{third interest.}
\end{array}
$$

595,50¾ the amount required. Ans. 595D. 50¾c.

5. What is the amount of 629 dols. for 7 years, at 6 per cent. per annum, compound interest?   Ans. 945 dols. 78 cts. 3m.

6. How much is the compound interest of 1256 dols. for 15 years, at 6 per cent. per annum ?   Ans. 1754 dols. 6 cts. 6m.

A TABLE shewing the amount of one pound or one dollar for any number of years under 33, at the rates of 5 and 6 per cent. per ann. compound interest.

Years.	5 Rates.	6	Years.	5 Rates.	6
1	1,05000	1,06000	17	2,29201	2,69277
2	1,10250	1,12360	18	2,40662	2,85434
3	1,15762	1,19101	19	2,52695	3,02559
4	1,21550	1,26247	20	2,65329	3,20713
5	1,27628	1,33822	21	2,78596	3,39956
6	1,34009	1,41852	22	2,92526	3,60353
7	1,40710	1,50363	23	3,07152	3,81975
8	1,47745	1,59384	24	3,22510	4,04893
9	1,55132	1,68948	25	3,38635	4,29187
10	1,62889	1,79084	26	3,55567	4,54938
11	1,71034	1,89829	27	3,73345	4,82234
12	1,79585	2,01219	28	3,92013	5,11168
13	1,88565	2,13292	29	4,11618	5,41838
14	1,97993	2,26090	30	4,32194	5,74349
15	2,07892	2,39655	31	4,53804	6,08810
16	2,18287	2,54035	32	4,76494	6,45338

The use of this Table is plain and easy, for multiplying the figures standing against the number of years, by the given principal, the product is the amount required.

COMPOUND INTEREST.

Examples.

7. What is the amount of 500 dollars, for 3 years, at 6 per cent. compound interest?

   1,19101 the tabular number for the time.
    500 the principal.
   ―――――
  595,50500

        Ans. 595 dols. 50 cts.

8. A merchant, on inspecting some old accounts in March, 1799, finds a settlement dated March, 1771, by which it appears there is due from him to A. B. £.2 8s. this sum he pays with compound interest at 6 per cent. per annum. The amount of it is required?

  5,11168 the tabular number for 28 years.
   2,4 the principal with the shillings inserted decimally.
 ―――――
 2044672
 1022336
―――――
£.12,268032
   20
―――――
s. 5,360640
   12
―――――
d. 4,327680
   4
―――――
qrs. 1,310720 Ans. £.12 5s. 4¼d. or 40 dols. 89 cts. 3 ms.

   Calculated in Federal Money.
   5,11168
     8 dollars.
  ―――――
 dols. 40,89344

   Ans. 40 dols. 89 cts. 3 mills, as above.

COMMISSION AND BROKERAGE.

Commission and Brokerage are compensations to Factors and Brokers for their respective services.

The method of operation is the same as in Simple Interest.

EXAMPLES.

1. What is the commission on £.596 18 4, at 6 per cent.?

```
  596 18 4      Or thus, £.5 | 1/20 |  596 18  4
        6                                 ─────
  ─────────               1 | 1/5  |   29 16 11
  35|81 10 0                            5 19  4½
     20                                ─────────
  ─────                               £.35 16  3½
  16|80
     12
  ─────
   3|60
      4
  ─────
   2|40                    Ans. £.35 16 3½
```

2. What is the commission on 1974 dollars at 5 per cent.?

```
  1974
     5
  ─────
  98,70              Ans. 98 dols. 70 cts.
```

3. What is the commission on £.526 11 5 at 3½ per cent.? Ans. £.18 8 7

4. What is the commission on £.1258 17 3 at 7⅔ per cent.? Ans. £.93 3 1¼.

5. What is the commission on 2176 dols. 50 cents, at 2½ per cent.? Ans. 54 dols. 41 cts. 2 m.

6. The sales of certain goods amount to 1873 dols. 40 cts. what sum is to be received for them, allowing 2½ per cent. for commission, and ¼ per cent. for prompt payment of the neat proceeds? Ans. 1821 dols. 99 cts. 9 m.

L

COMMISSION AND BROKERAGE.

7. Required the neat proceeds of certain goods amounting to £.456 11 8, allowing a commission of 2½ per cent.

£.5 $\frac{1}{20}$ | 456 11 8

2½ ½ . | 22 16 7 commission at 5 per cent.

11 8 3½ commission at 2½ per cent.

Ans. £.445 3 4½ neat proceeds.

8. What is the commission on £.1371 9 5 at 5 per cent.?
Ans. £.68 11 5½

9. What is the commission on £.1958 at 5½ per cent.?
Ans. £.107 13 9½

10. What is the commission on £.1859 7 6 at ⅞ per cent.?
Ans. £.16 5 4½

11. What is the brokerage on 1853 dols. at ¾ per cent.?
Ans. 13 dols. 89 cts. 7 m.

12. What is the brokerage on £.874 15 3 at ¼ per cent.?
Ans. £.2 3 8¾

13. What is the brokerage on 1298 dols. 53 cts. at ⅜ per cent.?

1298,53
3
———
8)3895,59
———
Dols. 4,86,94   Ans. 4 dols. 86 cts. 9 m.

14. What is the brokerage on £.1321 11 4 at 1⅛ per cent.?
Ans. £.14 17 4

15. A factor receives 988 dollars to lay out, after having deducted his commission of 4 per. cent. how much will remain to be laid out?

d.
100
4
———
       d.   d.
If 104 : 100 :: 988 : 950 dols. the answer.

16. A factor has in his hands 3690 dollars, which he is directed to lay out in iron, reserving from it his commission of 2½ per cent. on the purchase; the iron being 95 dols. per ton: how much did he purchase?
Ans. 37 tons 17 cwt. 3 qrs. 16$\frac{4}{19}$ lb.

INSURANCE.

INSURANCE is an exemption from hazard, by paying, or otherwise securing a certain sum, on condition of being indemnified for loss or damage.

Policy is the name given to the instrument, by which the contract of indemnity is effected between the insurer and insured.

Average loss is 5 per cent.; that is, if the insured suffer any loss or damage not exceeding 5 per cent. he bears it himself, and the insurers are free.

RULE. The method of operation as in interest.

EXAMPLES.

1. What is the premium of insuring £.924 at 7 per cent.?
Ans. £.64 13 7

2. What is the premium on 1650 dollars, at 12 per cent.?
Ans. 198 dols.

3. What is the premium of insuring 1250 dollars, at $7\frac{1}{2}$ per cent.? Ans. 93 dols. 75 cts.

4. What is the premium of insuring 4500 dollars, at 25 per cent.? Ans. 1125 dols.

5. What is the premium of insuring 1650 dollars, at $15\frac{1}{2}$ per cent.? Ans. 255 dols. 75 cts.

6. What is the premium of insuring 1873 dollars, at $\frac{1}{8}$ per cent.? Ans. 2 dols. 34 cts. 1 m.

7. What sum is to be received for a policy of 1658 dols. deducting the premium of 23 per cent.? Ans. 1276 dols. 66 cts.

8. What sum must a policy be taken out for to cover 1800 dollars, when the premium is 10 per cent.?

```
   100 Policy.
    10 Premium.
   ___
                    d.   d.    d.
    90 sum covered. If 90 : 100 :: 1800   Ans. 2000 dols.
```

Proof, 2000 dollars at 10 per cent.
```
       10
      ___
    200,00           the policy   2000
                     the premium   200
                                  ____
                     sum covered  1800 dols.
```

9. What sum must a policy be taken out for to cover 3926 dols. 7 cts. when the premium is 6 per cent.?
Ans. 4176 dols. 67 cts.

GENERAL AVERAGE.

WHATEVER the master of a ship in distress, with the advice of his officers and sailors, deliberately resolves to do, for the preservation of the whole, in cutting away masts or cables, or in throwing goods overboard to lighten his vessel, which is what is meant by jettison or jetson, is in all places, permitted to be brought into a general average, in which all, who are concerned in ship, freight and cargo, are to bear an equal or proportionable part of the loss of what was so sacrificed for the common welfare; and it must be made good by the insurers in such proportions as they have underwritten.

EXAMPLES
OF ADJUSTED AVERAGES.

1. A loaded ship met with such bad weather, that the master and mariners found it impossible to save her without throwing part of her cargo overboard, which they are authorized to do for preservation. Being thus necessitated, they threw such goods as lay nearest at hand, and lightened the ship of 10 casks of hardware, and 40 pipes of Madeira wine, which they judged sufficient to keep her from sinking. Soon after that the ship arrived at her destined port, and then an average bill was immediately made in order to adjust the loss, and to pay the proprietors of those goods, which were thrown overboard, for the good of the whole.

Average accrued to ship——, for goods thrown overboard for preservation of ship, freight and cargo.

	Dols.
Ship valued at	12000
Freight (wages and victuals deducted)	3000
Thomas Nugent's value of goods	4000
Thomas Morgan's value of goods	2500
James Simpson's value of goods	8500
Andrew Wilson for 40 pipes of wine	4000
Laurence Ward for 10 casks of hard ware	6000
	40000

	Dols.
Mr. Andrew Wilson's goods thrown overboard were valued at	4000
Mr. Laurence Ward .... do	6000
	10000

If 40000 give 10000 loss, what loss will 100 give?
          Ans. 25 per cent.

GENERAL AVERAGE.

The ship must pay to A. W. and L. W. for 12000
dollars, at 25 per cent. 3000
The freight 3000 dollars, at the same rate 750
Thomas Nugent, for 4000 dollars, at the same rate 1000
Thomas Morgan, for 2500 dollars, at the same rate 625
James Simpson, for 8500 dollars, at the same rate 2125

A. W. and L. W. receive of the owners of the goods saved,
and the ship's owners 7500
Their property being insured, the underwriters pay them 2500

10000

2. The Sea Horse, capt. Dix, laden with hemp, cordage, and iron, bound from Riga to Boston, ran on shore, coming through the grounds at Elsineur. The captain hired a great number of men, and several lighters, to lighten the ship, and to get her afloat again, which was done; but he was obliged to pay 409 dols. 23 cts. for their assistance. This expense being incurred to preserve both ship and cargo, the average must consequently be general. When the ship arrived at Boston, the captain immediately made a protest and an average bill, which was thus stated:

Average accruing to the ship Sea-Horse from Riga to Boston, in 1799, for assistance in getting off the strand of Elsineur.

                                                    dols. cts.
For sundry charges paid at the Sound for lighters and
  assistance in getting off the ship 409 23
Protest and postage 35 37
                                                 ───────
                                                 444 60

The ship's freight money 3460
Wages for all the people, 4 ms. and 20 d. 560 ⎫
Victuals for ditto .... 300 ⎭   860
                                                 ───────
                                                 2600

The ship Sea-Horse valued at 12000
Freight valued at 2600
William Jenkins for value of hemp 18000
Daniel Jones for value of cordage 4000
Enoch Flinn for value of iron 2400
                                              ───────
                                              39000

L 2.

If 39000 dols. lose 444 dols. 60 cts. what will 100 dols. lose?
Ans. 1 dol. 14 cts.

	dols.	cts.
The ship must bear 12000 dols. at 114 cts. per 100 dols.	136	80
The freight 2600 dols. at the same rate	29	64
William Jenkins for 18000	205	20
Daniel Jones for 4000	45	60
Enoch Flinn for 2400	27	36
	444	60

BUYING AND SELLING STOCKS.

STOCK, in the sense in which it is here used, is a fund established by government or individuals in a corporate capacity, the value of which is variable.

EXAMPLES.

1. What is the amount of 1565 dollars national bank stock, at 134 per cent.?

```
    1565
     134
    ----
    6260
    4695
    1565
    ------
   2097,10        Ans. 2097 dols. 10 cts.
```

2. What is the amount of 2958 dols. bank stock, at 25 per cent. advance?

```
           2958
   25   ¼   739,50
           ------
          3697,50     Ans. 3697 dols. 50 cts.
```

	dols.			dols. cts.
3.	6959	of 8 per cent. stock, at 110 per cent.	Ans.	7654,90
4.	1796	6 .... 91½		1643,34
5.	1284	3 .... 54¼		696,57
6.	3172	deferred 89		2823,08
7.	1518	state notes 83¾		1271,32½
8.	1686	Union Bank 128		2158,08

DISCOUNT

Is the abating of so much money to be received before it is due, as that money, if put at interest, would gain in the same time and at the same rate.

Thus 100 dollars would discharge a debt of 106 dollars payable in 12 months, discount at 6 per cent. per annum, because the 100 dollars received would, if put to interest, regain the 6 dollars discount.

RULE. As 100 dollars, with the interest for the given time, is to 100, so is the given sum to the present worth, and the difference between the present worth and the given sum is the discount.

EXAMPLES.

1. What is the present worth of 450 dols. due in 6 months, discount at 6 per cent. per annum?

$$6m. \quad \tfrac{1}{2} \quad 6$$
$$\overline{3}$$
$$100$$
$$\overline{} \quad d.$$
$$103 : 100 :: 450$$

Ans. 436 dols. 89 cts.

2. How much is the discount of £.308 15s. due in 18 months, at 8 per cent. per annum? Ans. £.33 1 7¾

3. What is the present worth of 5150 dols. due in 4½ months, discounting at the rate of 8 per cent. per annum, and allowing 1 per cent. for prompt payment? Ans. 4950 dols.

4. A is to pay 5927 dols. on the 19th of April, 1799, and 6989 dols. the 19th of July following—It is required to know how much money will discharge both sums on the 19th of January, 1799, discounting at 8 per cent. per annum? Ans. 11569 dols. 43 cts.

Though the discount found by the preceding method is thought to be the sum that should be deducted for present payment in justice to both parties, yet in business the interest for the time is taken for the discount.

DISCOUNT.

EXAMPLES.

5. What ready money will discharge a note of 150 dollars, due in 60 days, allowing legal interest, or 6 per cent. per annum as discount?

$$\frac{150}{1,50} \quad 1 = \text{half the months.}$$

150 the debt:
1,50 the interest:
───────
148,50   Ans. 148 dols. 50 cts.

6. Bought goods to the amount of 950 dollars, at 90 days credit, what ready money will discharge it, allowing the interest for the time at 6 per cent. per annum as discount?

Ans. 935 dols. 75 cts. if calculated for 3 months.
935 dols. 95 cts. if calculated for 90 days.

When the interest for the time is allowed as discount, it is presumed that neither party suffers any loss, but the following statement evinces the contrary.

A owes B 100 dollars payable in 12 months, for present payment of which B allows 6 dollars or the interest for the time, and receives 94 dollars; this sum he immediately lends to C for the same space of time, and then receives the amount, being 99 dollars 64 cents, which is 36 cents less than he would have to receive from A, had he left the money in his hands—but if he had allowed A the discount, and not the interest, for the time, he would have received from him 94 dols. 34 cents, and this sum being put to interest, would amount to 100 dols. in one year, which shews that the discount and not the interest, is the just deduction for prompt payment.

But when discount is to be made for present payment, without any regard to time, the interest of the sum, as calculated for a year, is the discount.

DISCOUNT.

EXAMPLES.

7. How much is the discount of 853 dols. at 2 per cent.?

$$853 \\ \cdot 2 \\ \overline{dols.\ 17,06}$$

Ans. 17 dols. 6 cts.

8. How much money is to be received for 985 dols. 75 cts. discounting 4 per cent.?   Ans. 946 dols. 32 cts.

BANK DISCOUNT.

THE method used among bankers, in discounting notes, &c. is, to find the interest of the sum, from the date of the note to the time when it becomes due, including the days of grace; the interest thus found is reckoned the discount. Thus, if a note for 100 dollars, dated the 2d September, be discounted at a bank, for 30 days, the interest of that sum for 33 days being 55 cents, is deducted for discount. It may be asked, why interest for 33 days is calculated on a note for 30, the answer is, that as custom has allowed the borrower three days of grace—that is, though the time of the note expires on the 1st of October (the day of the date being included in the 30 days) he may withhold the payment till the 4th—it is therefore reasonable that he should pay interest for it.

If a note of 100 dollars were discounted at a bank for 60 days, the interest of that sum for 63 days, being 105 cents, would be deducted for the same reason.

In case payment of a note be not convenient at the proper time, a new note must be presented on the day of discount, immediately preceding the expiration of the time, paying the same discount or interest for the time, as before stated. Thus, a note of 100 dollars, dated October 7th, 1800, for 30 days, though it is not payable till November 8th, yet must be replaced by a new note on Tuesday, November 4th, paying at the same time 55 cents. A note of the same date, for 100 dols. for 60 days, though not payable till Monday, December 8th, (including in this time the days of grace) must be replaced by a new note on Tuesday, December 2d, paying likewise 105 cents. In the former case the borrower sustains a loss of 5 days in 30, and in the latter 7 days in 60 by renewing. All Banks have their stated times of discount, generally once in a week. In the preceding cases, the Bank is supposed to discount on Tuesday. Some Banks discount twice a week—others oftener.

The discount of any sum, discounted for 30 or 60 days, is found by multiplying it by one sixth of the days. [See *interest*, page 110.]

EXAMPLES.

1. How much is the interest of 238 dols. discounted for 30 days?

```
   238
  ,5½ = ⅙ of 33 days.
  ─────
  1190
   119
  ─────
  1,30,9
```
Ans. 1 dol. 30 cts. 9 m.

2. What is the interest of 564 dols. discounted for 60 days?

```
   564
  1,0½ = ⅙ of 63 days.
  ─────
  5640
   282
  ─────
  5,92,2
```
Ans. 5 dols. 92 cts. 2 ms.

What is the discount of the following sums, viz.

	dols.		dols. cts. ms.
3.	159	discounted for 30 days. Ans.	0 87 4
4.	273	do.	1 50 1
5.	683	do.	3 75 6
6.	789	do.	4 33 9
7.	2194	do.	12 06 7
8.	219	discounted for 60 days. Ans.	2 29 9
9.	187	do.	1 96 3
10.	319	do.	3 34 9
11.	658	do.	6 90 9
12.	2169	do.	22 77 4

DISCOUNT. 131

13. How much is the discount of a debenture of 319 dols. payable in 210 days, discounting for 30 days.

NOTE. 28 days are allowed for a month, interest being calculated as if the note were renewable.

```
28)210(7 mo.              319
   196                     ,5½ = ¼ of 33 days.
   ———                    ————
   14 days.                159 5
                            15 9
                           ————
                           1,75,4 for 1 month.
                               7
                           ————
                           12,27,8 for 7 months.
         14 d. ½ mo.          877
                           ————
                           13,15,5
                           Ans. 13 dols. 15 cts. 5 m.
```

14. What is the discount of the above sum, discounting for 60 days?

NOTE. As notes are renewable in 56 days, the interest of all securities is calculated accordingly.

```
56)210(3 discount months.   319
   168                      1,0½ = ⅓ of 63 days.
   ———                      ————
   42 days.                  3190
                              159
                            ————
                             3,34,9 for 1 discount mo.
                                 3
                            ————
                            10,04,7 for 3 ditto.
            28d. ½ mo.       1,67,4
            14   ⅓             83,7
                            ————
                            12,55,8
                            Ans. 12 dols. 55 cts. 8 m.
```

The preceding examples shew the difference between discounting for 30 and 60 days.

DISCOUNT.

What is the discount of the following sums, discounting for 30 days?

	dols.	days.	dols. cts. ms.
15.	187 for	79	Ans. 2 90 0
16.	219	115	4 94 5
17.	658	47	6 7 4
18.	2169	128	54 53 2

What is the discount of the following sums, discounting for 60 days?

	dols.	days.	dols. cts. ms.
19.	187 for	79	Ans. 2 76 8
20.	219	115	4 72 2
21.	658	47	5 79 8
22.	2169	128	52 5 4

When a note is offered at a bank for discount, two endorsers are generally required, to the first of whom it is said to be payable: Thus—A having occasion for a sum of money, procures B and C as endorsers to his note, and offers it for discount in the following form:

100 *Dollars.* ———, ———

For value received, I promise to pay B, or order, at the ——— *Bank, on demand, one hundred dollars, with interest after* ——— *days.* A.

When state notes, bank shares, &c. are lodged in a bank as security for monies, a note is presented in this form:

For value received, I promise to pay the President, Directors and Company of the ——— *Bank, or their order, at said Bank, on demand,* ——— *dollars, with interest after* ——— *days.* C.D.

EQUATION OF PAYMENTS.

THE design of this Rule is to find a mean time for the payment of several sums due at different times.

RULE. Multiply each sum by its time, and divide the sum of the products by the whole debt; the quotient is accounted the mean time.

EQUATION OF PAYMENTS.

EXAMPLES.

1. A owes B 200 dols. whereof 40 dols. is to be paid in 3 months, 60 dols. in 5 months, and the remainder in 10 months, at what time may the whole be paid without any injustice to either?

```
        dols.    mo.
         40  ×  3 =  120
         60  ×  5 =  300
        100  × 10 = 1000
        ───              
        200      200)1420
                 ───────
                 7 months and 3 days.
```

2. A is indebted to B £.120, whereof one half is to be paid in 3 months, one quarter in 6 months, and the remainder in 9 months, what is the equated time for the payment of the whole? Ans. 5 months and 7½ days.

3. C owes D 1400 dols. to be paid in 3 months, but D being in want of money, C pays him, at the expiration of 2 months, 1000 dols. how much longer than 3 months ought C, in equity, to defer the payment of the rest? Ans. 2½ months.

Those who are exact in these calculations, find the present worth of each particular sum, then find on what time these present worths will be increased to the total of the particular sums payable at the particular times to come; and that is the true equated time for the payment of the whole.

......

BARTER

Is the exchanging of one commodity for another on such terms as may be agreed on.

EXAMPLES.

1. How many quintals of fish, at 2 dols. per quintal will pay for 140 hhds. of salt, at 4 dols. 70 cts. per hhd.?

```
            140
            4,70
            ────
           9800
            560
   dols.  qtl.  ────
  If  2  :  1  ::  658,00 the amount of the salt.
       ────
          Ans.  329 quintals.
```

M

2. A buys of B 4 hhds. of rum containing 410 gallons, at 1 dol. 17 cts. per gallon ; and 253 lb. of coffee, at 21 cts. per lb. in part of which he pays 21 dollars in cash, and the balance in boards, at 4 dols. per thousand ; how many feet of boards did the balance require ? Ans. 127957½ feet.

3. B has C's note for 250 dols. with 6 months interest due on it, and to redeem it C delivers him 60 bushels of wheat at 7s. 6d. per bushel, 50 bushels of corn at 5s. 3d. per bushel, and the balance in staves at 30 dols. per thousand ; what number of staves did B receive ?

Ans. 5550 staves, or 4 m. 6 hun. and 10 casts.

4. B bought of D the hull of a schooner of 70 tons, at 16 dols. per ton, and paid him in cash 500 dols. 3 hhds. of molasses containing 350 gallons, at 64 cts. and is to pay the balance in New-England rum at 74 cts. per gallon ; how many gallons is D to receive ? Ans. 535$\frac{5}{37}$ gals.

5. A buys of B 250 quintals of fish, at 25s. per quintal ; in payment B takes 100 dols. in cash, 2 hhds. of molasses containing 87 and 92 gals. at 3s. 8d. per gallon, 1 pipe of brandy containing 120 gals. at 7s. 6d. per gallon, and gives 3 months credit for the remainder ; required the balance due, and what cash would pay it, allowing the interest of it for the time at 6 per cent. per annum, as discount for prompt payment ?

Ans. Balance is 682 dols. 27 cts. 6 ms.=672,04,2 in cash.

6. C sells to D 28,674 feet of boards at 8 dols. 50 cts. per thousand, and takes in payment ⅓ cash, 4 barrels N. E. rum containing 128 gallons at 78 cts. per gallon, 1 barrel of sugar weighing neat 2 cwt. 2 qrs. 4 lb. at 10 dollars per cwt. and the balance in coffee at 25 cts. per lb. ; how much money and coffee is C to receive ?

Ans. 81 dols. 24 cts. 3 ms. and 149$\frac{39}{250}$ lb. of Coffee.

7. C has nutmegs worth 7s. 6d. per lb. in ready money, but in barter he will have 8s. ; D has tobacco worth 9d. per lb. ; how much must he rate it per lb. that his profit may be equal to C's ? Ans. 9$\frac{3}{5}$d.

8. A has tea which he barters with B at 10d. per lb. more than it cost him, against cambrick which stands B in 10s. per yard, but he puts it at 12s. 6d. ; I would know the first cost of the tea ? Ans. 3s. 4d. per lb.

9. A has 240 bushels of rye, which cost him 90 cts. per bushel ; this he barters with B at 95 cts. for wheat that stands B in 99 cts. per bushel ; how many bushels of wheat is he to

receive in barter, and at what price is it to be rated, that their gains may be equal?

Ans. $218\frac{38}{209}$ bushels, at $104\frac{1}{2}$ cts. per bushel.

10. A and B barter some goods—A put his at $30\frac{6}{25}$ shillings, and gains 8 per cent. B puts his at $24\frac{3}{10}$ shillings, and gains at the same rate; what was the first cost of the goods?

Ans. 28s. and 22s. 6d.

11. A and B barter; A has cloth that cost 28d. B's cost him 22d. and he puts it at 25d.; how high must A put his to gain 10 per cent. more than B? Ans. 35d.

12. C and D barter—C makes of 7s. 6s. 8d. D makes of 7s. 6d. 7s. 3d.; who has lost most, and by how much per cent.?

Ans. C loses $1\frac{1}{4}$ per cent. more than D.

• • • • • •

LOSS AND GAIN

Is a rule that discovers what is gained or lost in buying or selling goods, and instructs merchants and traders to raise or fall the price of their goods so as to gain or lose so much per cent. &c.

EXAMPLES.

1. Bought a piece of broadcloth containing 53 yards, at 4 dols. 65 cts. per yard, and sold at 5 dols. per yard; what is the profit on the whole?

```
              dols.cts.
                 5
                4,65
      yd.      ─────      yds.
  If   1  :  ,35   ::     53
                          35
                         ────
                         265
                         159
                       ──────
                        18,55   Ans. 18 dols. 55 cts.
```

2. If 1 lb. of coffee cost 28 cts. and is sold for 31 cts. what is the profit on 3 bags, weighing 293 lbs. neat?

Ans. 8 dols. 79 cts.

LOSS AND GAIN.

3. Bought a piece of baize of 42 yards, for £.4 14 6, and sold it at 2s. 6d. per yard; what is the gain or loss on the whole piece ? Ans. 10s. 6d. gain.

4. A merchant bought 59 cwt. 3 qr. 14 lb. of iron, at 112 dols. per ton, paid freight and charges, 24 dols. what is the gain or loss, if he sells the whole at 37s. 4d. per cwt. ?
Ans. 13 dols. 26 cts. gain.

5. If a gallon of wine cost 6s. 8d. and is sold for 7s. 2d. what is the gain per cent. ?

```
            7  2
            6  8
      s. d. ───        £.
   If 6  8  :  6  ::  100   Ans. 7½ per cent. gain.
```

6. When 20 per cent. loss is made on coffee, sold at 20 cts. per lb. what was the first cost ? Ans. 25 cts.

7. At 13½ cts. profit on the dollar, how much is it per cent. ?
Ans. 13½ per cent. or 13 dols. 50 cts. per 100 dols.

8. A trader sells his goods at 2½d. profit on the shilling, how much is it per cent. ? Ans. 20⅚, or £.20 16 8

9. Which is the better bargain, in purchasing fish, 17 shillings per quintal, and 4 months credit, or 16s. 8d. cash ?
Ans. They are alike.

PROOF. The present worth of 17s. found by discount, is equal to 16s. 8d. and 16s. 8d. with 4 months interest, will amount to 17s.

10. A bought a piece of shalloon, containing 34 yards, at 3s. 4d. per yard, and sold it at 12½ per cent. loss, how much did he sell it per yard ? Ans. 2s. 11d.

11. Bought rum at 90 cts. per gallon, at what rate must it be sold to gain 20 per cent. ? Ans. 108 cents.

12. A trader bought 1 hhd. of rum, of a certain proof, containing 115 gallons, at 1 dol. 10 cts. per gallon, how many gallons of water must he put into it to gain 5 dollars, by selling it at 1 dollar per gallon ? Ans. 16½ gallons.

13. Bought 4 hhds. of rum, containing 450 gallons, at 1 dol. per gallon, and sold it at 1 dol. 20 cts. per gallon, and gave 3 months credit ; now allowing the leakage of the rum while in my possession to be 10 gallons, I would know the gain or loss, discounting for the present worth of the debt at 6 per cent. per annum ? Ans. 70 dols. 19 cts. gain.

14. A vintner buys 596 gallons of wine, at 6s. 3d. per gallon, in ready money, and sells it immediately at 6s. 9d. per gallon, payable in 3 months, how much is his gain or loss, supposing he allows the interest for the time, at 6 per cent. per annum, as discount for present payment ? Ans. £.11 17 8 gained.

15. What would be the gain or loss on the aforesaid wine, supposing the discount for present payment to be made at 2 per cent. without any regard to time ? Ans. £.10 17 6½ gain.

16. A merchant bought a parcel of cloth at the rate of 1 dol. for every 2 yds. of which he sold a certain quantity at the rate of 3 dols. for every 5 yds. and then found he had gained as much as 18 yards cost, how many yards did he sell ? Ans. 90 yds.

17. Bought rum at 1 dol. 25 cts. per gallon, which not proving so good as I expected, I am content to lose 18 per cent. by it, how must I sell it per gallon ? Ans. 1 dol. 2½ cts.

18. H sells a quantity of corn at 1 dollar a bushel, and gains 20 per cent. some time after he sold of the same, to the amount of 37 dols. 50 cts. and gained 50 per cent. how many bushels were there in the last parcel, and at what rate did he sell it per bushel ?· Ans. 30 bushels, at 1 dol. 25 cts. per bushel.

19. A distiller is about purchasing 10000 gallons of molasses, which he can have at 48 cents per gallon, in ready money, or 50 cents with 2 months credit, it is required to know which is more advantageous to him, either to buy it on credit, or to borrow the money at 8 per cent. per annum to pay the cash price ?· Ans. He will gain 136 dols. by paying the cash.

20. A tobacconist buys 4 hogsheads of tobacco weighing 38 cwt. 2 qrs. 8 lb. gross, tare 94 lb. per hhd. at 9 dols. per cwt. ready money, and sells it at 11½d. per lb. allowing tare at 14 lb. per cwt. to receive two-thirds in cash, and for the remainder a note at 90 days credit ; his gain or loss is required, supposing the note is discounted at a bank where discount is made for 60 days. Ans. 285 dols. 43 cts. gain.

ALLIGATION MEDIAL

Is when the quantities and prices of several things are given, to find the mean price of the mixture compounded of those things.

RULE. As the sum of the quantities or whole composition is to their total value, so is any part of the composition to its mean price.

EXAMPLES.

1. A grocer would mix 25 lb. of raisins, at 8 cents per lb. and 35 lb. at 10 cents per lb. with 40 lb. at 12 cents per lb.— what is 1 lb. of this mixture worth?

```
     lb.           cts.         cts.
     25     at      8  ....    200
     35     ......  10  ....   350
     40     ......  12  ....   480
     ───                       ────
     100                       1030
     lb.          cts.          lb.
 If  100   :     1030    ::     1
```

1|00)10|30

cts. 10,3   Ans. 10 cents, 3 mills.

2. A goldsmith mixes 8 lb. 5½ oz. of gold, of 14 carats fine, with 12 lb. 8½ oz. of 18 carats fine; what is the fineness of this mixture?   Ans. $16\frac{54}{113}$ carats.

3. A grocer would mix 12 cwt. of sugar, at 10 dols. per cwt. with 3 cwt. at 8⅔ dols. per cwt. and 8 cwt. at 7½ dols. per cwt. what will 5 cwt. of this mixture be worth?
Ans. 44 dols. 78 cts. 2 ms.

4. A refiner melts 2½ lb. of gold, of 20 carats fine, with 4 lb. of 18 carats fine; how much alloy must he put to it to make it 22 carats fine?
Ans. It is not fine enough by $3\frac{3}{13}$ carats, so that no alloy must be put to it, but more gold.

5. A malster mingles 30 quarters of brown malt, at 28s. per quarter, with 46 quarters of pale, at 30s. per quarter, and 24 quarters of high dried ditto, at 25s. per quarter; what is the value of 8 bushels of this mixture? Ans. £.1 8s. 2½d. ⅗

ALLIGATION MEDIAL.

6. If I mix 27 bushels of wheat, at 5s. 6d. the bushel, with the same quantity of rye, at 4s. per bushel, and 14 bushels of barley, at 2s. 8d. per bushel, what is the worth of a bushel of this mixture? Ans. 4s. 3¾d. 28/68

7. A grocer mingled 3 cwt. of sugar, at 56s. per cwt. 6 cwt. at £.1 17 4 per cwt. and 3 cwt. at £.3 14 8 per cwt. what is 1 cwt. of this mixture worth? Ans. £.2 11 4

8. A mealman has flour of several sorts, and would mix 3 bushels at 3s. 5d. per bushel, 4 bushels at 5s. 6d. per bushel, and 5 bushels at 4s. 8d. per bushel, what is the worth of a bushel of this mixture? Ans. 4s. 7½d. 4/12

9. A vintner mixes 20 gallons of Port, at 5s. 4d. per gallon, with 12 gallons of White wine, at 5s. per gallon, 30 gallons of Lisbon, at 6s. per gallon, and 20 gallons of Mountain, at 4s. 6d. per gallon, what is a gallon of this mixture worth? Ans. 5s. 3¾d. 50/82

10. A farmer mingled 20 bushels of wheat, at 5s. per bushel, and 36 bushels of rye, at 3s. per bushel, with 40 bushels of barley, at 2s. per bushel, I desire to know the worth of a bushel of this mixture? Ans. 3 shillings.

11. A person mixing a quantity of oats, at 2s. 6d. per bushel, with the like quantity of beans, at 4s. 6d. per bushel, would be glad to know the value of 1 bushel of that mixture? Ans. 3s. 6d.

12. A refiner having 12 lb. of silver bullion of 6 oz. fine, would melt it with 8 lb. of 7 oz. fine, and 10 lb. of 8 oz. fine, required the fineness of 1 lb. of that mixture? Ans. 6 oz. 18 dwt. 16 grs.

13. If with 40 bushels of corn, at 4s. per bushel, there are mixed 10 bushels, at 6s. per bushel, 30 bushels, at 5s. per bushel, and 20 bushels, at 3s. per bushel, what will 10 bushels of that mixture be worth? Ans. £.2 3s.

ALLIGATION ALTERNATE

Is the method of finding what quantity of any number of simples, whose rates are given, will compose a mixture of a given rate; so that it is the reverse of Alligation Medial, and may be proved by it.

ALLIGATION ALTERNATE.

RULE. 1. Write the rates of the simples in a column under each other.

2. Connect or link with a continued line the rate of each simple which is less than that of the compound, with one, or any number, of those that are greater than the compound, and each greater rate with one or any number of the less.

3. Write the difference between the mixture rate and that of each of the simples, opposite the rates with which they are linked.

4. Then if only one difference stand against any rate, it will be the quantity belonging to that rate ; but if there be several, their sum will be the quantity.

EXAMPLES.

1. A merchant would mix wines at 14s. 19s. 15s. and 22s. per gallon, so that the mixture may be worth 18s. the gallon ; what quantity of each must be taken ?

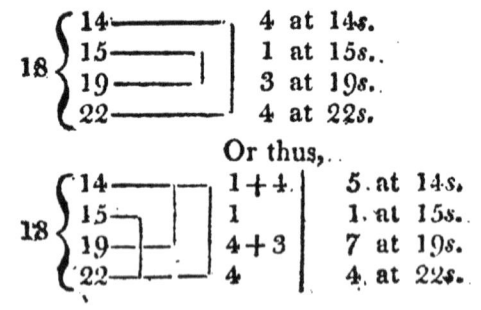

$$18 \begin{cases} 14 \\ 15 \\ 19 \\ 22 \end{cases} \begin{array}{l} 4 \\ 1 \\ 3 \\ 4 \end{array} \begin{array}{l} \text{at } 14s. \\ \text{at } 15s. \\ \text{at } 19s. \\ \text{at } 22s. \end{array}$$

Or thus,

$$18 \begin{cases} 14 \\ 15 \\ 19 \\ 22 \end{cases} \begin{array}{l} 1+4 \\ 1 \\ 4+3 \\ 4 \end{array} \begin{array}{l} 5 \text{ at } 14s. \\ 1 \text{ at } 15s. \\ 7 \text{ at } 19s. \\ 4 \text{ at } 22s. \end{array}$$

NOTE. Questions in this rule admit of a great variety of answers, according to the manner of linking them.

2. How much wine, at 6s. per gallon, and at 4s. per gallon, must be mixed together, that the composition may be worth 5s. per gallon ?   Ans. 1 qt. or 1 gall. of each, &c.

3. How much corn, at 2s. 6d. 3s. 8d. 4s. and 4s. 8d. per bushel, must be mixed together, that the compound may be worth 3s. 10d. per bushel ?
   Ans. 12 at 2s. 6d. 12 at 3s. 8d. 18 at 4s. and 18 at 4s. 8d.

4. A goldsmith has gold of 17, 18, 22 and 24 carats fine ; how much must he take of each to make it 21 carats fine ?
   Ans. 3 of 17, 1 of 18, 3 of 22, and 4 of 24.

ALLIGATION ALTERNATE.

5. It is required to mix brandy at 8s. wine at 7s. cider at 1s. and water together, so that the mixture may be worth 5s. per gallon?

   Ans. 9 gals. of brandy, 9 of wine, 5 of cider, and 5 of water.

When the whole composition is limited to a certain quantity.

RULE. Find an answer as before by linking; then say, As the sum of the quantities, or differences thus determined, is to the given quantity, so is each ingredient, found by linking, to the required quantity of each.

EXAMPLES.

6. How many gallons of water must be mixed with wine worth 3s. per gallon, so as to fill a vessel of 100 gallons, and that a gallon may be afforded at 2s. 6d.?

$$30\begin{cases} 0\text{———}6 \\ 36\text{———}30 \end{cases}$$
$$\overline{36}$$

```
36 : 100 :: 6           36 : 100 :: 30
         6                       30
      ─────                    ─────
   36)600(16                36)3000(83
      36                       288
      ───                      ───
      240                      120
      216                      108
      ───                      ───
       24                       12
```

  Ans. $83\frac{1}{3}$ gallons of wine, and $16\frac{2}{3}$ of water.

7. A grocer has currants at 4d. 6d. 9d. and 11d. per lb. and he would make a mixture of 240 lb. so that it might be afforded at 8d. per lb. how much of each sort must he take?

   Ans. 72 lb. at 4d. 24 at 6d. 48 at 9d. and 96 at 11d.

8. How much gold of 15, of 17, of 18, and of 22 carats fine, must be mixed together, to form a composition of 40 oz. of 20 carats fine?

   Ans. 5 oz. of 15, of 17, and of 18, and 25 oz. of 22.

ALLIGATION ALTERNATE.

When one of the ingredients is limited to a certain quantity.

RULE. Take the difference between each price and the mean rate, as before ; then,

As the difference of that simple, whose quantity is given, is to the rest of the differences severally, so is the quantity given, to the several quantities required.

EXAMPLES.

9. How much wine, at 5s. at 5s. 6d. and at 6s. the gallon, must be mixed with three gallons, at 4s. per gallon, so that the mixture may be worth 5s. 4d. per gallon ?

$$64\begin{cases}48\\60\\66\\72\end{cases}\quad \begin{array}{l}8+2=10\\8+2=10\\16+4=20\\16+4=20\end{array}$$

10 : 10 :: 3 : 3
10 : 20 :: 3 : 6
10 : 20 :: 3 : 6

Ans. 3 gallons at 5s. ; 6 at 5s. 6d. and 6 at 6s.

10. A grocer would mix teas at 12s. 10s. and 6s. with 20 lb. at 4s. per lb. ; how much of each sort must he take to make the composition worth 8s. per lb. ?

Ans. 20 lb. at 4s. ; 10 lb. at 6s. ; 10 lb. at 10s. ; and 20 lb. at 12s.

11. How much gold of 15, of 17, and of 22 carats fine, must be mixed with 5 oz. of 18 carats fine, so that the composition may be 20 carats fine ?

Ans. 5 oz. of 15 carats fine, 5 oz. of 17, and 25 of 22.

POSITION.

POSITION is a rule, which, by false or supposed numbers, taken at pleasure, discovers the true one required. It is divided into two parts, SINGLE and DOUBLE.

SINGLE POSITION

Is, by using one supposed number, and working with it as the true one, you find the real number required by the following,

POSITION. 143

RULE. As the total of the errors is to the given sum, so is the supposed number to the true one required.

PROOF. Add the several parts of the result together, and if it agrees with the given sum, it is right.

EXAMPLES.

1. A school-master, being asked how many scholars he had, said, If I had as many, half as many, and one quarter as many more, I should have 264; how many had he?

Suppose he had 72
As many 72
½ as many 36
¼ as many 18

As 198 : 264 :: 72
                72

                         Proof.
        528   96
       1848   96
                          48
198)19008(96 Answer.   24
       1782   ——
       ——   264
       1188
       1188

2. A person, after spending ⅓ and ¼ of his money, had 60 dollars left; what had he at first? Ans. 144 dols.

3. A certain sum of money is to be divided between 4 persons, in such a manner, that the first shall have ½ of it, the second ¼, the third ⅙, and the fourth the remainder, which is 28 dollars; what was the sum? Ans. 112 dols.

4. A person lent his friend a sum of money unknown, to receive interest for the same, at 6 per cent. per annum, simple interest, and at the end of 5 years he received for principal and interest 644 dollars 80 cents; what was the sum lent? Ans. 496 dols.

DOUBLE POSITION

Is, by making use of two supposed numbers, which, if both prove false, are, with their errors, to be thus disposed:

RULE. 1. Place each error against its respective position.
2. Multiply them cross wise.

POSITION.

3. If the errors are alike, that is, both greater or both less than the given number, divide the difference of the products by the difference of the errors, and the quotient is the answer: But if the errors be unlike, divide the sum of the products by the sum of the errors, and the quotient will be the answer.

EXAMPLES.

1. B asked C how much his horse cost; C answered, that if he cost him three times as much as he did, and 15 dollars more, he would stand him in 300 dollars; what was the price of the horse?

```
     dols.                  dols.
Suppose he cost 90     Suppose he cost 96
                3                      3
              ───                    ───
              270                    288
               15                     15
              ───                    ───
       285 too lit. by 15 dls.  303 too much by 3 dls.
```

$$\begin{array}{cc} 90 & 15- \\ \times & \\ 96 & 3+ \end{array}$$

```
    15   1440    270
     3    270
   ───   ────
Sum of the errors 18 ) 1710 ( 95 answer      95
                        162                   3
                       ───                  ───
                         90                 285
                         90                  15
                       ───                  ───
                                         300 proof.
```

2. Two persons, A and B, have both the same income; A saves one-fifth of his yearly: but B, by spending 150 dollars per annum, more than A, at the end of 8 years finds himself 400 dollars in debt; what is their income, and what does each spend per annum?

Ans. Their income is 500 dollars per annum; also A spends 400, and B 550 dollars per annum.

3. There is a fish whose head is 9 inches long, and his tail is as long as his head and half his body, and his body is as long as the head and tail; what is the whole length of the fish?

Ans. 6 feet.

4. Divide 15 into two such parts, so that when the greater is multiplied by 4, and the less by 16, the products will be equal. Ans. 12 and 3.

5. A man had two silver cups of unequal weight, having one cover to both, 5oz.; now if the cover is put on the less cup it will be double the weight of the greater cup, and put on the greater cup it will be three times as heavy as the less cup: what is the weight of each cup? Ans. 3 oz. less—4 oz. greater.

6. A person being asked, in the afternoon, what o'clock it was, answered that the time past from noon was equal to $\frac{2}{13}$ of the time to midnight; required the time?
Ans. 36 minutes past one.

* * * * *

EXCHANGE.

EXCHANGE is the paying of money in one place or country, for the like value to be received in another place or country.

There are two kinds of money, viz. Real, and Imaginary.

Real money is a piece of metal coined by the authority of the State, and current at a certain price, by virtue of the said authority, or of its own intrinsic value.

Imaginary money is a denomination used to express a sum of money of which there is no real species, as a *livre* in France, a *pound* in America, because there is no species current, in this or that country, precisely the value of either of the sums.

Par of Exchange is the intrinsic value of the money of one country compared with that of another country, as one pound sterling is equal to thirty-five shillings Flemish.

Course of Exchange is the current or running price of exchange, which is sometimes above, and sometimes below par, varying according to the occurrences of trade, or demand for money. Of this course, there are tables published daily in commercial cities: thus by Lloyd's List, of 3d. December, 1799, the course of exchange between Hamburgh and London, was 32s. 6½d. Flemish, per pound sterling, being 2s. 5½d. under par, or loss to London.

EXCHANGE.

GREAT-BRITAIN.

The money of account is pounds, shillings, pence and farthings.

The English Guinea is 21 shillings, Sterling.

Weights and measures generally as in the United States.

The United States dollar is equal to 4s. 6d. Sterling.

To change Sterling to Federal money.

RULE. Annex three cyphers to the sum (if pounds only) and multiply it by 4; this product divide by 9, and you have the answer in cents. If there be shillings, &c. the usual method is to reduce it to Massachusetts money, by adding one third to it, and then reduce this sum to Federal.

EXAMPLES.

1. Change £.48 Sterling to Federal.
48000
4
―――――
9)192000
―――――
21333$\frac{1}{3}$ cents.   Ans. 213 dols. 33$\frac{1}{3}$ cts.

2. Change £.389 17 4$\frac{1}{2}$ Sterling to Federal, exchange at 33$\frac{1}{3}$ per cent. that is, £.133$\frac{1}{3}$ Massachusetts for £.100 Sterling.
$\frac{1}{3}$)389 17 4$\frac{1}{2}$ Sterling
129 19 1$\frac{1}{2}$ Exchange
―――――
519 16 6   Massachusetts
―――――
,3)519,825
―――――
Cts. 173275 Federal. Ans. 1732 dols. 75 cts.

NOTE. Sterling is changed to Massachusetts money by adding one-third to the sum, and Massachusetts to Sterling by deducting one-fourth from it.

・・・・・・

To change Federal Currency to Sterling.

RULE. Work by either of the following methods.

EXCHANGE.

EXAMPLES.

Change 1732 dollars 75 cents to sterling.

First Method.
            1732

4s.   ½   346 8
6d.   ⅛   43 6
50 cents   2 3
25 cents   1 1½
           ─────────
Ans.   £.389 17 4½

Second Method.
          1732,75
             ,3
          ────────
          519|825
             20
          ────────
          16|500
             12
          ────────
          6|000

¼)519 16 6 Massachusetts
   129 19 1½ Exchange
  ─────────────────
Ans. £.389 17 4½ Sterling.

1. What is the Federal amount of an invoice of goods, charged at £.196 14 6 Sterling advancing on it 25 per cent?

        25   ¼)196 14 6 Sterling
             49 3 7½ Advance
             ─────────────
            245 18 1½
Exchange at 33⅓ per cent. 81 19 4½
                         ─────────────
                         £.327 17 6   Massachusetts

3)327875
─────────
cts. 109291⅔   Ans. 1092 dols. 91⅔ cts.

2. The Sterling cost of certain goods being £.60 12 6, what does it amount to in Massachusetts money, advancing on it 50 per cent.?

                          60 12 6
50 per cent. advance   30 6 3
                         ─────────
                          90 18 9
Exchange at 33⅓ per cent. 30 6 3
                         ─────────
         Ans.   £.121 5 0 Massachusetts money.

The mercantile method, with 50 per cent. advance, is to double the Sterling for Massachusetts money; thus,
           60 12 6 Sterling.
              2
          ─────────
          £.121 5 0 Massachusetts, as above.

EXCHANGE.

3. An invoice of goods, charged at £.52 19 7 sterling, is sold at 75 per cent. advance on the sterling cost, how much is it in Massachusetts money?

```
                          52 19  7
      Advance at 50       26  9  9½
                 25       13  4 10¾
                          ─────────
                          92 14  3¼
Exchange at 33⅓ per cent. 30 18  1
                          ─────────
              Ans. £.123 12 4¼ Massachusetts money.
```

The mercantile method, with 75 per cent. advance, is to multiply the sterling by 2⅓ for Massachusetts money.

```
Thus,   52 19  7
                2⅓
             ─────────
           105 19  2
            17 13  2¼
             ─────────
           £.123 12 4¼ Massachusetts money, as above.
```

4. The sterling cost of certain goods being £.214 11 6, how much is it in Federal money, advancing thereon 60 per cent?

```
              214 11  6
   50 ½       107  5  9  ⎫
   10 ⅕        21  9  1¾ ⎬ advance
              ─────────   ⎭
              343  6  4¾
Exchange ⅓    114  8  9½
              ─────────
              457 15  2¼  Massachusetts

Or thus,      214 11  6  Sterling
Exchange ⅓     71 10  6
              ─────────
              286  2  0
   50 ½       143  1  0
   10 ⅕        28 12  2¼
              ─────────
              457 15  2¼  Massachusetts.

          ,3)457,759
          ─────────
   Dollars 1525,86⅓   Ans. 1525 dols. 86⅓ cts.
```

5. What is the amount of a bill of exchange of £.115 14 9 sterling, sold in Boston at 1½ per cent. advance?

⅓)115	14	9	Sterling	
38	11	7	Exchange	
154	6	4	Massachusetts money	

,3)154,317

514,39 Federal
1½

51439
25719

Cents 771|58

	dols.	cts.
Value at par	514	39
Advance	7	71½
Amount	522	10½

Or thus,

	dols.	cts.		
Value at par	514	39		
Adv. at 1 pr ct.	5	14	3	
½ do.	2	57	1	
	7	71	4	Adv. at 1½ pr ct.
Amount	522	10	4	

6. A merchant in Boston receives a parcel of goods from London, charged in the invoice at the following prices, and marks them for sale at 60 per cent. advance on the sterling cost; required the selling price of each in Massachusetts money?

s.	d.		s.	d.		dols.	c.	m.
13	8	sterling, adv. 60 per ct.	29	1½	Massa. money, or	4	85	3
5	10	....	12	5¼	....	2	7	3
3	4	....	7	1¼	....	1	18	3
6	1½	....	13	0¼	....	2	17	6
17	0	....	36	3	....	6	4	
33	1	....	70	6½	....	11	75	6
1	2	....	2	5½	....		41	
18	10	....	40	2	....	6	69	4
11		....	23	5½	....	3	91	
2	4	....	4	11¼	....		82	3
32	3	....	68	9½	....	11	46	6
27	9	....	59	2¼	....	9	86	3

N 2

7. A watch that cost 15 guineas in London, was sold in Boston at 50 per cent. advance on the sterling cost, what was the price?

15 guineas=£.15 15 0 Sterling
2

31 10 0 Massachusetts.

,3)31,5

Ans. 105 dollars.

8. How much is the premium of insuring £.294 at 8 guineas per cent.? Ans. £.24 13 11 Sterling.

* * * * * *

Mercantile methods of calculating, viz.

At 25 per ct. disc. from the sterling cost, multiply it by 1 for the answer in Massachusetts money.

10	....	$1\frac{1}{5}$
par	....	$1\frac{1}{3}$
$12\frac{1}{2}$	per ct. adv. on the sterling cost, multiply it by	$1\frac{1}{2}$
25	....	$1\frac{2}{3}$
$31\frac{1}{4}$	....	$1\frac{3}{4}$
50	....	2
$62\frac{1}{2}$	....	$2\frac{1}{6}$
65	....	$2\frac{1}{5}$
75	....	$2\frac{1}{3}$
$87\frac{1}{2}$	....	$2\frac{1}{2}$
100	....	$2\frac{2}{3}$
125	....	3
140	....	$3\frac{1}{5}$
150	....	$3\frac{1}{3}$
$162\frac{1}{2}$	....	$3\frac{1}{2}$
175	....	$3\frac{2}{3}$
200	....	4

IRELAND.

The money of account as in England, but different in value. The par between England and Ireland is $8\frac{1}{3}$ per cent. that is, £.100 sterling money is £.108 6 8 in Ireland.

Mercantile weights and measures, the same as in England.

The United States dollar is equal to 4*s.* $10\frac{1}{2}$*d.* Irish.

The English guinea is equal to 22*s.* 9*d.* Irish.

To reduce Irish money to Federal.

Rule. Reduce the given sum to half pence, annex two cyphers to it, and then divide by 117, (the half pence in a dollar)

EXCHANGE. 151

and the quotient is the answer in cents. Or reduce the Irish to Sterling, by deducting $\frac{1}{13}$ from it, and then work as for Sterling.

EXAMPLE.

Change £.278 15 9 Irish money to Federal.

```
     First method.              Second method.
       278 15  9           ¹⁄₁₃)278  15   9 Irish
        20                       21   8  11 Exchange
       ─────                     ─────────────
       5575                     257   6  10 Sterling
         12                      85  15   7¼
       ─────                     ─────────────
       66909                    343   2   5¼ Mass.
          2                      ─────────────
       ─────                     ,3)343,122
     9)13381800                  ─────────────
9×13=117 ─────                   1143,74 cents
      13)1486866
       ─────
       114374 cents          Ans. 1143 dols. 74 cts.
```

To change Federal money to Irish.

RULE. Multiply the given sum by 117, reject two figures from the product to the right hand, and the remaining figures are the half pence in the given sum.

1. Change 1143 dols. 74 cts. to Irish.

```
           114374
              117
          ─────────
           800618
           114374
           114374
          ─────────
        2)133817|58
          ─────────
       12)66908¾
          ─────────
        2|0)5571|5  8
          ─────────
       Ans. £.278  15  8¾
```

If the sum is dollars only, work by either of the following methods.

EXCHANGE.

2. Change 1537 dollars to Irish.

First method.	Second method.
1537 at 4s. 10½d.	1537
	,3

```
4s.  ½   307  8
8d.  ⅙    51  4  8            461  2     Massachusetts
2    ¼    12 16  2   ¼        115  5  6  Exchange at 25 per ct.
½    ¼     3  4  0½           ─────────
         ─────────            345 16  6  Sterling
Ans. £374 12 10½   1/12        28 16  4½ Ex.8⅓pr.ct. or 1d. on 1s.
                              ─────────
                              £.374 12 10½
```

In changing Sterling to Irish money at par, $\frac{1}{12}$ is added to the sum for Irish; and in changing Irish to Sterling, $\frac{1}{13}$ is deducted for Sterling because 12 pence English are equal to 13 pence Irish, making the Exchange 1d. in a shilling, 1s. 8d. in a pound, and £.8 6 8 per cent.

EXAMPLES.

1. Change £.394 17 6 Sterling to Irish, at par, or £.8⅓ per cent.

```
  1/12 )394 17  6
         32 18  1½
        ─────────
```
Ans. £.427 15 7½ Irish.

2. Change £.427 15 7½ Irish money to Sterling, at 8⅓ per cent. in favour of England.

```
  1/13 )427 15  7½
         32 18  1½
        ─────────
```
Ans. £.394 17 6 Sterling.

3. Change £.370 Sterling to Irish, at 9 per cent.

```
    £.      £.     £.
   100  :  109  ::  370     Ans. £.403 6 0
```

4. Reduce £.403 6 Irish money to Sterling, at 9 per cent.

```
    9
  100
  ───                £.   s.
  109  :  100  ::  403   6     Ans. £.370
```

EXCHANGE. 153

HAMBURGH.

Accounts are kept in Hamburgh in Marks, Shillings Lubs or Stivers, and Deniers.

12 deniers, or 2 grotes, make ····1 shilling lubs, or stiver.
16 shillings lubs, stivers, or ⎫
32 grotes ············· ⎬ 1 mark.

ALSO,
12 grotes or pence Flemish make 1 shilling Flemish.
20 shillings Flemish ·········· 1 pound.

NOTE. 3 marks ········· make ······ 1 rix dollar.
7½ do. ················· 1 pound Flemish.
A shippound in Hamburgh ···· 280 lb.
A ring of staves ·· do ········· 240
100 lb. in Hamburgh ········ 107¼ lb. in U. States.
100 ells. ·· do. ············ 62½ yards.

The currency of Hamburgh is inferior to the bank money; the *agio*, or rate, is variable; May 14th, 1798, it was 20 per cent. in favour of the bank.

The mark banco is 33⅓ cents; (See laws of the U. States.)

EXAMPLES.

1. Change 12843 marks to Federal, at 33⅓ cts. per mark.
33⅓ = ⅓)12843

Ans. 4281 dollars.

2. In 4967 marks 8 stivers banco, how many dollars, exchange as above?
33⅓ = ⅓)4967,

$$1655,66⅓$$
8 stivers   ,16½

Dols. 1655,83

Ans. 1655 dols. 83 cts.

To change Hamburgh money to Sterling.

RULE. As the given rate is to one pound, so is the Hamburgh sum to the Sterling required.

154   EXCHANGE.

Examples.

1. Change 2443 marks 9½ stivers to Sterling, exchange at 32s. 6d. Flemish per pound Sterling.

```
   s.  d.       £.        m.      st.
  32   6   :   1   ::    2443     9½
  12 grotes.               32      2
  ─────                   ─────
   390                   4886   19 grotes.
                         7329
                           19
                         ─────
                         78195
                             1
                         ─────
                    390)78195(200£.
                         780
                         ───
                          195
                           20
                         ─────
                    390)3900(10s.
                         3900
                         ────
```

Ans. £.200 10 0

2. In 12093 marks 12 stivers, how many pounds sterling, exchange at 32s. 3d. Flemish per pound Sterling?
Ans. £.1000

3. In 4178 marks 2 stivers, how many pounds Sterling, exchange at 31s. 10d. Flemish per pound Sterling?
Ans. £.350

4. Change 1971 marks 13 stivers to Sterling, exchange at 35s. 6d. Flemish per pound Sterling.   Ans. £.148 2 4

· · · ·

To change Sterling to Hamburgh money.

RULE. As 1 pound Sterling is to the given rate, so is the Sterling sum to the Hamburgh required.

EXCHANGE.

EXAMPLE.

Change £.350 Sterling to Hamburgh money, exchange at 31s. 10d. Flemish per pound Sterling.

```
 £.    s.  d.         £.
 1  :  31  10   ::   350
       12
       ---
      382 grotes
      350
      ----
    19100
     1146
     -----
  2)133700 grotes
    ------
 16)66850 stivers
    ------
    4178 2         Ans. 4178 marks 2 stivers.
```

Proving the answers in the preceding case will further exemplify this.

To reduce Current to Bank money.

RULE. As 100 marks with the agio added, is to 100 bank, so is the current money to the bank required.

EXAMPLES.

1. Change 560 marks 8 stivers current to banco, agio at 18 per cent.

```
         18
        100
        ---
    118 : 100 :: 560 8. Ans. 475 marks.
```

2. Change 2366 marks current to banco, agio at 20 per cent.   Ans. 1971 marks, 10¾ stivers.

3. Change 7456 current marks to banco, agio at 22 per cent.   Ans. 6111 marks, 7 stivers.

EXCHANGE.

To change Bank to current money.

RULE. As 100 marks is to 100 with the agio added, so is the bank given to the current required.

EXAMPLES.

1. Change 475 marks banco to current, agio at 18 per ct.

```
          18
         100
   m.    ———    m.
  100 : 118 :: 475        Ans. 560 marks, 8 stivers.
```

Or thus,
```
    475
     18            475   bank
    ———            85 8  agio
    3800          ————
     475          560 8  as above.
    ————
    85|50
      16
    ————
     8|00
```

2. Change 1971 marks, 10⅔ stivers banco to current, agio at 20 per cent.

```
              m.      s.
    20 ⅕)1971    10⅔  banco
            394    5⅓  agio
         ————————————
    Ans. 2366     0   current.
```

· · · · · ·

PRACTICAL QUESTIONS.

1. How much will 63452 lb. of cotton come to, at 8 grotes per lb.?

```
      lb.   gr.    lb.
       1  :  8  ::  63452
                       8
                   ————————
               2)507616 grotes
               ————————
              16)253808 stivers
               ————————
         Ans.   15863 marks.
```

EXCHANGE. 157

2. What will 351 lb. of cotton come to at 50d. per lb. ?

NOTE. *d.* is the mark for pence Flemish, equal in value to half stivers or half shillings lubs.

```
   lb.     d.         lb.
    1  :  50   : :   351
                      50
          ─────────────────
          2)17550 grotes or pence flemish.
          ─────────────
          16)8775 stivers.
          ─────────────
               548  7   Ans. 548 marks 7 stivers.
```

3. What will 339 bars Russian iron come to, wt. 19662 lb. at 35½ marks per shippound ?

```
   lb.      m.         lb.
   280  :  35½  : :  19662      Ans. 2492 m. 14 stiv.
```

			m.	st.
4.	280 lb. of cotton ······ at	21 grotes per lb. ·····	183	12
5.	4002¼ lb. coffee ··········	8¼ stivers ············	2063	10
6.	2438 pipe staves ··········	16 marks per ring of 240 ··	162	9
7.	3540 hhd. ditto ···········	8½ ditto ditto ······	125	6
8.	529 barrel ditto ··········	5¼ ditto ditto ······	11	9
9.	1790 lb. sugar ············	21¼ pence per lb. ·····	1188	10
10.	4892 lb. rice ·············	18¼ marks per 100 ·····	892	12
11.	4 pieces 10-4 bedtick ···	24 ditto ·············	96	0
12.	140 half pint tumblers ····	8 ditto per 100 ·······	11	3
13.	100 boxes window glass ···	23 ditto per box ······	2300	
14.	1526¼ lb. coffee ··········	16¼ stivers per lb. ···	1574	3
15.	245 bars iron, wt. 8434 lb. ··	41 marks per shippound ····	1235	
16.	10 bales hemp, wt. 14108 lb.	74 ditto ditto ······	3728	

17. What is the commission on 18270 marks, at 2½ per cent. ?
Ans. 456 m. 12 st.

18. What is the interest of 6370 marks, for 3 months, at 5 per cent. per annum ? Ans. 79 m. 10 st.

EXCHANGE.

19. Change 5955 marks 7½ stivers to Dutch florins, at 38½ grotes per florin.

```
                          mar.    st.
                          5955    7½
grotes in a mark =  32     2  grotes a stiver.
                        ─────
                         11910  15 grotes in 7½ stivers.
                         17865
                            15
                        ─────
grotes  38½             190575 grotes.
         2                   2
                      ─────────
                77 ) 381150 ( 4950 guilders.
                     308
                     ───
                      731
                      693
                     ───
                      385
                      385
                     ───
```
          Ans. 4950 gild. or flor.

20. An American merchant orders his correspondent in Amsterdam to remit 4980 florins 16½ stivers to Hamburgh; this being done, when the exchange is 39¼ stivers for 2 marks, what sum is he credited for in Hamburgh?

```
  st.      M.      F.      st.
  39¼   :  2   ::  4980    16½
   4                 20
 ─────              ─────
  157               99616½
                        2
                    ─────
                    199233
                        4
                    ─────
          157)796932(5076 marks.
              785
              ────
              1193
              1099
              ────
               942
               942
              ────
```
          Ans. 5076 marks.

EXCHANGE.

HOLLAND.

Accounts are kept in Florins or Gilders, Stivers, Deniers or Pennings.

  8 pennings ············make········ 1 grote.
  2 grotes, or 16 pennings ············ 1 stiver.
  20 stivers, or 40 grotes ············· 1 gilder or florin.

ALSO,

  12 grotes, or 6 stivers ··············· 1 shilling.
  20 shillings, or 6 gilders ············· 1 pound Flemish.
  2½ florins····························· 1 rix dollar.

The florin or gilder of the United Netherlands is estimated in the United States at 40 cents, or 2 cents per stiver.

  100 lb. in Amsterdam make 109½ lb. in the U. States.
  100 ells ····do.··········· 75 yards do.

In liquid measure, 16 mingles make 1 steckan, 8 steckans 1 aum.

1. Change 1954 florins to Federal money, at 40 cts. per florin.

    1954
     40
    ――――
  dols. 781,60   Ans. 781 dols. 60 cts.

2. Change 2653 gilders 17 stivers to Federal money, at 40 cents per gilder.

  2653 17 Or thus, 2653 17
    40 2   20
  ―――― ――   ――――
  106120 34   53077 stivers.
     34   2 cts. per stiver.
  ――――   ――――
  106154 cts.   1061,54
         Ans. 1061 dols. 54 cts.

3. Change 1061 dols. 54 cts. to gilders, at 40 cts. per gilder.

   2)106154 cents.
   ――――――――
   2|0)53077|7 stivers.
    ―――――――
    2653 17 Ans. 2653 gild. 17 stiv.

160 EXCHANGE.

3. What must be paid in Boston for an invoice of goods charged at 591 florins 17 stivers; allowing the exchange at 40 cents per florin, or 2 cts. per stiver, and advancing on it 60 per cent. ?

```
    591 17
     20
   ─────
   11837 stivers.
       2
   ─────
dols. 236,74
     60 per cent.
   ─────
   142,0440
```

```
                    d.   c.
Am. of invoice,   236  74
Advance,          142  04
                 ─────────
            Ans. 378  78.
```

....

To change Sterling to Flemish.

RULE. As 1 pound sterling is to the given rate, so is the sterling given to the Flemish required.

EXAMPLES.

1. In £.100 10s. sterling, how many gilders, exchange at 33s. 9d. Flemish per pound sterling?

```
 £.        s.  d.      £.    s.
 1    :   33  9   ::  100   10.
 20       12           20
 ──      ───          ────
 20      405 grts.    2010
                       405
                      ────
                      10050
                      80400
                     ─────────
                   4)81405,0
                     ─────────
                     2)10702½ grotes.
                     2|0)2035|1¼ stivers.
                        ──────
                        1017 11¼     Ans. 1017 gild. 11¼ st.
```

....

To change Flemish to Sterling.

RULE. As the given rate is to £.1 sterling, so is the Flemish given to the sterling required.

EXCHANGE.

EXAMPLE.

Change 1017 gilders 11¼ stivers to sterling, exchange at 33s. 9d. Flemish per £. sterling.

```
 s.  d.       £.         fl.    st.
33  9   :   1    ::    1017   11¼
12                       40     2
―――                    ――――
405 grotes.            40680   22½
                          22½
```

405)40702½(100
    405
    ―――
    202½
     20

405)4050(10
    4050   Ans. £.100 10

To change Current Money to Bank.

RULE. As 100 gilders with the agio added, is to 100 bank, so is the current money given to the bank required.

EXAMPLE.

Change 823 gilders 9½ stivers current money into bank, agio at 4½ per cent.

```
  g.           g.          g.    s.
104½   :    100   ::    823    9½
 20                      20
―――                    ―――――
2090                   16469½
                         100
```

2090)1646920(788 gilders.

To change Bank Money into Current.

RULE. As 100 gilders bank is to 100 with the agio added, so is the bank money given to the current required.

EXAMPLE.

Change 788 gilders bank money to current, agio at 4½ per cent.

    g.   g.   g.
  100 : 104½ :: 788   Ans. 823 gilders, 9½ stiv.

EXCHANGE.

PRACTICAL QUESTIONS.

1. What will 1867 lb. of coffee come to at 19 stivers per lb.?

```
    1867
      19
   -----
   16803
    1867
   -----
2|0)3547|3 stivers.
   -----
   1773 13       Ans. 1773 gilders, 13 stivers.
```

2. What will 92 hhds. of sugar come to, weighing 104242 lb. gross, deducting 2 per cent. for good weight, tare 18 per cent. at 21 grotes per lb. ?

```
                     104242
deduct 2 per cent.     2085
                     ------
                     102157
tare 18 per cent.     18388
                     ------
                      83769 nt. wt.
                         21
                     ------
                      83769
                     167538
                     ------
                2)1759149 grotes.
                     ------
                2|0)87957|4½ stivers.
                     ------
                     43978 14½    Ans. 43978 gilders, 14½ stivers.
```

3. What will 251 bars of iron come to, weighing gross 10364 lb. at 9¾ gilders per 100 lb. deducting 2 per cent. ?

```
   10364
      9¾
   -----
   93276
    5182
    2591
   -----
   1010,49                         g.   s.  p.
       20             2 pr. ct. = 1/50)1010  9  12
   -----                                20   4   3
    9,80                                ----------
      16                         Ans.  990   5   9
   -----
   12,80
```

EXCHANGE. 163

4. What will 143 steckans 2 mingles of brandy come to, at 42 gilders per aum?

```
        8)143
           17  7  2
           42
           ──────
           34
           68
4 steckans  ½   21
2 ......    ¼   10 10
1 ......    ⅛    5  5
2 mingles   1/16  0 13  2
           ──────────
          751  8  2    Ans. 751 gild. 8 stiv. 2 pennings.
```

				gild.	st.
5.	21315 lb. of sugar	23 grotes per lb.		12256	2
6.	56560	25		35350	
7.	27093	25¼		17271	15
8.	8189 lb. coffee	23¼ stivers		9622	1
9.	4650	23¼		5405	12
10.	1970	19¾		1945	7
11.	39285	21¼		41740	6
12.	212 ells linen, 208 payable	30		312	
13.	4190 lb. butter	13 gild. per 40 lb.		1361	15
14.	6476	11¼		1861	17
15.	2012 lb. lead	13½ do. per 100 lb.		271	12
16.	214 steck. 11 ming. brandy	42 do. per aum.		1127	3

DENMARK.

Accounts are kept in Danish current dollars and skillings, reckoning 96 skillings to the dollar.

The course of exchange on London in September, 1799, was 5 rix or Danish dollars for 1 pound sterling.

The rix dollar of Denmark is estimated at 100 cents.—(See Laws of the United States.)

96 pounds of Denmark make 100 pounds in the U. States. Their weights are shippounds, lispounds and pounds—

16 pounds ...... make ...... 1 lispound.
20 lispounds, or 320 pounds ...... 1 shippound.

1. How much will 8 pieces of platillas come to, at 9 dols. 56 skills. per piece?

```
        9  56
            8
       ──────
       76  64        Ans. 76 dols. 64 skills.
```

EXCHANGE.

2. How much will 1418 bars of iron come to, weighing 263 shippounds 9 lispounds and 4 pounds, at 15 dols. per shippound?

```
    lb.    d.    s.  lis. lb.           Or,      ship.
   320 :  15  :: 263  9   4                      263
                  20                              15
                 ────                            ────
                 5269                    lis.    3945
                   16                    5.  ¼   3 72
                 ────                    4   ⅓   3 00
                .31618                   4lb. ¹⁄₁₆  0 18
                 5269                           ─────
                 ────                    Ans.   3951 90
                 84308
                   15
                 ────
     32|0)126462|0(3951
          96
          ───
          304
          288
          ───
           166
           160
           ───
             62
             32
            ───
             30
             96
            ───
     32)2880(90
        2880
        ────
```
Ans. 3951 dols. 90 sk.

3. What is the commission on 21545 Danish dols. 13 skills. at 2 per cent.?

```
        21545. 13
                2
        ─────────
        430,90  26
             96
        ─────
          566
          810
        ─────
          86,66
```
Ans. 430 dols. 86 skills.

4. What will 4 hhds. of sugar come to, weighing gross 4314 lb. tare 17 per cent. at 22 skillings per lb.?

Ans. 820 dols. 62 skills.

		dls. sks.		dls. sks.
5.	4 pieces table cloth	3 80		15 32
6.	50	9 56		479 16
7.	13	17 64		229 64
8.	24	12		288 00
9.	50	15		750 00
10.	100 coils cord, wt. 52sh. 16l 2lb.	30 per shippound		1884 18
11.	85 bun. cl hemp, 250	36		9000 00
12.	1951 bars Rus. iron, 362 8 10	14		5074 3

13. How many Danish dollars will be received in Copenhagen, for a bill of £.2300 on London, exchange at 5 rix dollars per pound sterling? Ans. 11500 dols.

14. A bill is drawn in Copenhagen for 18574 marks, 7 stivers, Hamburgh money, when the exchange is 128 Danish dollars for 100 rix dollars in Hamburgh, how many Danish dollars does it amount to?

NOTE. Three marks are equal to 1 rix dollar.

  m. r.d.   m. st.   r.d. sk.
If 3 : 1 :: 18574 7 : 6191 46.

  r.d. D.d.   r.d. sk.
If 100 : 128 :: 6191 46   Ans. 7925 Dan. dols. 6 sk.

Or thus,   3)18574 7 Hamburgh money.
            ─────────
            6191 46
28 per cent. 1733 56
            ─────────
            7925 6 Dan. money, as above.

BREMEN.

Accounts are kept in rix dollars and grotes, reckoning 72 grotes to the rix dollar, which is equal to 2¼ marks.

On the 29th Nov. 1795, the exchange on London was 551 rix dollars for £.100 sterling.

In 1802, the course of exchange on the United States was 75 cents per rix dollar.

The Bremen last is equal to 80 bushels in the U. States.

100 lb. in Bremen are equal to 110 lb. in the U. States.

EXCHANGE.

1. What will 1104 lb. of coffee come to at 32¾ grotes per lb.?

$$\begin{array}{r} 1104 \\ 32\frac{3}{4} \\ \hline 2208 \\ 3312 \\ 552 \\ 276 \\ \hline \end{array}$$

               r.d. grotes.
72)36156(502 12
    360
    ―――
    156
    144
    ―――
     12   Ans. 502 rix dols. 12 grotes.

2. What is the commission on 7621 rix dols. 6 gr. at 3½ per cent.?   Ans. 266 rix dols. 53 grotes.

                                            r. dols. gr.
3.   3071 lb. coffee · · 32¾ grotes per lb. · · 1396 63
4.   400 · · · · · · · · · 32⅝ · · · · · · · · · · · · · · 181 18
5.   706 · · · · · · · · · · 33½ · · · · · · · · · · · · · · 328 35
6. 31407 lb. sugar · · 15¾ · · · · · · · · · · · · · · 6870 20

ANTWERP.

Accounts are kept in Antwerp in gilders, shillings, and grotes.

    12 grotes · · · · · · · · · · make · · · · · · · · · · 1 shilling.
    3⅓ shillings, or 40 grotes · · · · · · · · · · 1 gilder.

The Braband or Antwerp grotes are of the value of the cents of the United States, a gilder being reckoned at 40 cents. In the current money of Antwerp they have stivers of the value of the stiver of Amsterdam, or 2 cents United States currency.

    100 pots Braband   =   86½ gallons U. States.
    96 lb. Antwerp   =   100 lb.   do.
    100 Braband ells, about   74 yds. American.

The new quintal of Antwerp consists of 10 myriagrammes or 204 lb. 14 oz. Avoirdupois weight.

The loss on sugar exported from America to Antwerp is 22½ per cent. viz. tare 14 lb. per 100 lb.—good weight 2 lb.—loss of weight 5 lb.—discount 1½ lb. equal to 22½ lb. per 100 lb.

Loss on cotton 12½ per cent.—on coffee in bags 11½ per cent.

EXCHANGE.

EXAMPLES.

1. A cargo consisting of 48 hhds. sugar, weighing 376 cwt. 1 qr. 14 lb. valued per invoice at 12 dols. per cwt. and 63 bags coffee weighing 7345 lb. at 32 cents per lb. is sold in Antwerp; what sum was received for it, in gilders and grotes, at 40 cents per gilder, allowing the customary deductions for tare, &c. at an advance of 33⅓ per cent. from the invoice?

	cwt. qr. lb.		lb.
	376 1 14		7345
Tare,&c. 22½ per ct.	84 2 20½	Tare,&c. 11½ per ct.	844½
Neat	291 2 22½	Neat	6500½
			32
	dols. cts.		13000
	12 00		19500
	10		16
	120 00	dols.	2080,16
	10		
	1200 00		
	2		
	2400 00	val. of 200 cwt.	
	1080 00	 90	
	12 00	 1	
	6 00	...... 2 qrs.	
	1 50	...... 14 lb.	
	75	... 7	
	10 7	. 1	
	5 3	. ½	
Value of sugar	3500 41 0	291 2 22½	
do. coffee	2080 16 0		
	5580 57 0	4)0)7440)6 cents.	
Adv. 33⅓ = ⅓	1860 19 0		
		18601 36	
Dols.	7440 76 0		

Ans. 18601 gild. 36 gr.

2. What sum must be paid in Boston for an invoice of goods imported from Antwerp, amounting to 7315 gilders, exchange 40 cents per gilder, at an advance of 40 per cent?

```
    7315                    7315
       40 per cent. adv.       2926 adv.
    ─────                   ─────
    2926,00                10241
                              40 cents per gild.
                           ─────
                           4096,40
```

Ans. 4096 dols. 40 cts.

....

RUSSIA.

Accounts are kept in Petersburgh, in Rubles and Copecs, reckoning 100 copecs to 1 ruble.

The course of exchange on London, in July, 1796, was $34\tfrac{3}{4}d.$ sterling per ruble.
Ditto .... on Amsterdam .... 30 stivers banco per ruble.
Ditto .... on Hamburgh, Aug. 1798, $22\tfrac{1}{2}$ st. banco do.
Ditto .... on U. States, Sept. 1802, 55 cents   do.

100 lb. Petersburgh weight are equal to $88\tfrac{3}{4}$ lb. in the U. States.

Their weights are Barquits, Poods, Pounds, and Zollotnicks—
96 zollotnicks .... make .... 1 pound.
40 pounds .... 1 pood.
10 poods .... 1 barquit.

Their long measure is the Arsheen, of 28 American inches: 9 arsheens are equal to 7 yards.

1. What will 7500 arsheens of ravens-duck come to, at $14\tfrac{1}{2}$ rubles for 50 arsheens?

```
arsh.    rub.      arsh.
 50  :  14½   ::  7500      Ans. 2175 rubles.
```

EXCHANGE.

2. What will 813 poods 5 lb. of clean hemp come to, at $30\frac{1}{2}$ rubles per barquit?

```
   lb.      rub.      p.  lb.
  400  :   30½  ::   813  5
                      40

                    32525
                      30½

                    975750
                     16262

               4|00)9920|12

                    2480,03
```
Ans. 2480 rubles 3 copecs.

3. What will 2846 poods 5 lb. of bar iron come to, at 200 copecs per pood?

```
                    2846
                     200

                  569200
    5 lb.  ⅝         25
    copecs       569225
```
Ans. 5692 rubles 25 copecs.

4. What is the commission on 5256 rub. 33 cop. at 3 per ct.?

```
         5256,33
               3

         157,68,99
```
Ans. 157 rubles 68 copecs.

			rub.	cop.
5.	4997½ arsheens flems	24 rubles per 50 arsheens.	2398	80
6.	1700 do. drillings	34 copecs per arsheen.	578	
7.	355 do. ticking	100 do. do.	355	
8.	118¾ do. do.	110 do. do.	130	62
9.	200 pieces of sail cloth	21 rubles per piece.	4200	
10.	2101 poods 25 lb. hemp	31 do. per barqnit.	6515	04

11. How many rubles must be received in Petersburgh for a bill of 15500 gilders on Amsterdam, when the exchange is 30 stivers per ruble?

```
   st.     cop.     gild.                    gild.
 As 30  :  100  ::  15500    Or thus,  ⅓)15500
                      20                 5166,66⅔

                  310000 stivers.         10333,33⅓
                     100

               3|0)3100000|0

                  10333,33⅓
```
Ans. 10333 rub. 33⅓ cop.

P

12. A bill of £.3000 Sterling is drawn on London, exchange at 31¾d. Sterling per Ruble, what is its value in Petersburgh?

```
        d.           rub.        £.
  As  31¾    :    1   ::    3000
       4                      20
      ───                   ─────
      127                   60000
                               12
                            ──────
                            720000
                                 4
                           ────────
              127)2880000(22677 rubles
                  254
                  ───
                  340
                  254
                  ───
                   860
                   762
                   ───
                    980
                    889
                    ───
                     910
                     889
                     ───
                 127)2100(16 copecs
                     127
                     ───
                      830
                      762     Ans. 22677 rub. 16 cop.
                      ───
                       68
```

Two cyphers are annexed to the remainder instead of multiplying by 100 copecs.

FRANCE.

12 deniers = 1 sol, 20 sols = 1 livre.

The crown of exchange is 3 livres tournois.

A livre tournois of France is estimated at 18½ cents in the United States.

Note. The word *tournois* is applied to the money of France, as sterling is to the money of England.

EXCHANGE.

1. Change £.1220 sterling to French money, exchange at 17⅝d. per crown of 3 livres tournois.

```
       d.      liv.      £.
      17⅝  :   3   ::   1220
       8                  20
      ───                ─────
      141               24400
                           12
                        ──────
                        292800
                             8
                       ───────
                       2342400
                             3
                       ───────
         141)7027200(49838 livres
             564
             ────
             1387
             1269
             ────
             1182
             1128
             ────
              540
              423
             ────
             1170
             1128
             ────
               42
               20
             ────
         141)840(5s.
             705
             ───
             135
              12
             ───
         141)1620(11d.
             141
             ───
              210
              141
             ───
               69      Ans. 49838 liv. 5 sol. 11 d
```

172 EXCHANGE.

2. Change £.400 sterling to French money, exchange at 17¾d. sterling per crown of 3 livres. Ans. 16225 liv. 7s.0¾ ⁶⁄₁d.

3. Change 4224 livres tournois to sterling, exchange at 17½d. per crown of 3 livres.

```
    liv.       d.       liv.
     3    :   17½  ::  4224
                       17½
                      ─────
                      29568
                       4224
                       2112
                      ─────
                   3)73920

                  12)24640

                  2|0)205|3 4
                      ─────
                      102 13 4
```

Ans. £.102 13s. 4d.

Or, Take ⅓ of the given sum to reduce it to crowns, and multiply by the rate of exchange; the product will be the answer in pence.

```
   ⅓)4224 livres
     ─────
     1408 crowns
       17½
     ─────
     9856
     1408
      704
     ─────
   12)24640 pence

   2|0)205|3 4
       ─────
       £.102 13 4  as above.
```

4. Change 49838 livres 5s. 11⅔¾d. to sterling, exchange at 17⅝d. sterling per crown. Ans. £.1220.

5. What will 2434 velts of brandy come to, at 320 livres per 29 velts? Ans. 26857 liv. 18s. 7d.

EXCHANGE.

6. What is the freight of 3302½ velts, at 9 livres per ton of 120 velts? Ans. 247 liv. 13s. 9d.

7. What is the commission on 36591 liv. 2s. 4 den. at 2½ per cent.? Ans. 914 liv. 15s. 6 den.

8. What is the interest of 66476 liv. 10s. 9 den. for 1 month and 10 days, at ½ per cent. per month?

```
         ½)66476 10  9
         ───────────
         332|38  5  4
              20
            ─────
            7|65
              12
            ─────
            7|84
         332    7  7
10 days ⅓ 110  15 10
         ───────────
Ans. Liv. 443  3  5
```

9. What is the interest of 3255 livres, for 28 days, at ½ per cent. per month?

```
      ½)3255
      ─────
      16|27 10
         20
      ─────
      5|50
        12
      ─────
      6|00
      16  5  6  for one month
15 days ½  8  2  9
10 .... ⅓  5  8  6
 3 .... ⅕  1 12  6
          ─────────
Ans. Liv. 15  3  9
```

The present money of account in France is in francs and centimes or hundredths.

In Nov. 1800, an English guinea was worth 25 fr. 75 cent. A Spanish dollar ...... 5 do. 53 do.

P 2

To change francs to livres tournois.

RULE. Multiply the francs by 81 and divide by 80 for livres.

EXAMPLE.

Change 3756 francs to livres.

```
       3756
         81
      ─────
       3756
      30048
      ──────
8,0)30423,6
      ──────
      3802 76
           20
         ────
       8,0)152,0
           ────
            19       Ans. 3802 liv. 19 sols.
```

To change livres tournois to francs.

RULE. Multiply the livres by 80, and divide the product by 81 for francs.

EXAMPLE.

Change 5469 livres to francs.

```
        5469
          80
       ──────
81)437520(5401,43
   405
   ───
    325
    324
   ───
     120
      81
    ────
     390
     324
    ────
     260
     243
    ────
      17      Ans. 5401 fr. 43 cen.
```

EXCHANGE.

To change sols and deniers to centimes.

RULE. Take one half of the sols and deniers, as if they were integers; this half is the number of centimes required.

EXAMPLES.

	sol. den.	sol. den.	sol. den.	sol. den.	
Change	4 6	12 2	6 8	16 6	to centimes.
Ans.	23	61	34	83	centimes.

When there is a remainder in dividing the sols, it is to be carried to the deniers, and reckoned 10 and not 12; add this 10 to the deniers, and take one half of the sum for the remaining centime.

EXAMPLES.

	sol. den.	sol. den.	sol. den.	
Reduce	5 8	15 4	19 6	to centimes.
Ans.	29	77	98	centimes.

If the number of deniers be 10 or 11, they are to be rejected, and in place of them you are to add 1 to the number of sols preceding, and then annex a cypher to it; one half of this is the centimes required.

EXAMPLES.

	sol. den.	sol. den.		sol. den.	
Change	1 10	7 11	and	15 10	to centimes.
	2)20	2)80		2)160	
Ans.	10	40		80	centimes.

Sols and deniers are reduced to centimes by the preceding rule, and though the result is not accurate, yet from its simplicity and conciseness it is generally used.

EXCHANGE.

TABLES

For changing Livres, Sols and Deniers to Francs and Centimes.

[N. B. The first is sufficiently exact for business; in the second the answer is calculated to the ten-thousandths part of a centime.]

	Tab. I.			Tab. II.	
	Fr.	Cent.	Fr.	Cent.	10,000ths of a centime.
Deniers.					
1	0	0	0	0	4115
2	0	1	0	0	8230
3	0	1	0	1	2346
4	0	2	0	1	6461
5	0	2	0	2	0576
6	0	2	0	2	4691
7	0	3	0	2	8807
8	0	3	0	3	2922
9	0	4	0	3	7037
10	0	4	0	4	1152
11	0	5	0	4	5267
Sols.					
1	0	5	0	4	9383
2	0	10	0	9	8765
3	0	15	0	14	8148
4	0	20	0	19	7531
5	0	25	0	24	6914
6	0	30	0	29	6296
7	0	35	0	34	5679
8	0	40	0	39	5062
9	0	44	0	44	4444
10	0	49	0	49	3827
11	0	54	0	54	3210
12	0	59	0	59	2593
13	0	64	0	64	1975
14	0	69	0	69	1358
15	0	74	0	74	0741
16	0	79	0	79	0123
17	0	84	0	83	9506
18	0	89	0	88	8889
19	0	94	0	93	8272
Livres.					
1	0	99	0	98	7654
2	1	98	1	97	5309
3	2	96	2	96	2963
4	3	95	3	95	0617
5	4	94	4	93	8272
6	5	93	5	92	5926
7	6	91	6	91	3580
8	7	90	7	90	1235
9	8	89	8	88	8889
10	9	88	9	87	6543

EXCHANGE.

Livres.	Fr. Cent.	Fr. Cent.	10,000ths of a centime.
12	11 85	11 85	1852
15	14 81	14 81	4815
20	19 75	19 75	3086
24	23 70	23 70	3704
30	29 63	29 62	9630
40	39 51	39 50	6173
50	49 38	49 38	2716
60	59 26	59 25	9259
70	69 14	69 13	5803
72	71 11	71 11	1111
80	79 01	79 01	2346
90	88 89	88 88	8889
95	94 81	94 81	4815
100	98 77	98 76	5432
200	197 53	197 53	0864
300	296 30	296 29	6297
400	395 06	395 06	1729
500	493 83	493 82	7161
1000	987 65	987 65	4322
5000	4938 27	4938 27	1608
10000	9876 54	9876 54	3217

A TABLE
FOR REDUCING FRANCS AND CENTIMES TO LIVRES, SOLS AND DENIERS.

Cent.	sol. den.	100ths of den.	Francs.	liv. sol. den.
1	0 2	43	2	2 0 6
2	0 4	86	3	3 0 9
3	0 7	29	4	4 1 0
4	0 9	72	5	5 1 3
5	1 0	15	6	6 1 6
10	2 0	30	7	7 1 9
15	3 0	45	8	8 2 0
20	4 0	60	9	9 2 3
25	5 0	75	10	10 2 6
30	6 0	90	15	15 3 9
35	7 1	05	20	20 5 0
40	8 1	20	30	30 7 6
45	9 1	35	40	40 10 0
50	10 1	50	50	50 12 6
55	11 1	65	60	60 15 0
60	12 1	80	70	70 17 6
65	13 1	95	80	81 0 0
70	14 2	10	90	91 2 6
75	15 2	25	100	101 5 0
80	16 2	40	200	202 10 0
85	17 2	55	300	303 15 0
90	18 2	70	400	405 0 0
95	19 2	85	500	506 5 0
			1000	1012 10 0
Francs.	liv. sol. den.		5000	5062 10 0
1	1 0 3		10000	10125 0 0

SPAIN.

SPANISH reckonings are of two sorts—
Money of plate, distinguished *hard* or *plate* dollars, &c.
Money of vellon, distinguished by *current* dollars.
The former is 88 4/7 per cent. above the latter.
100 reals plate being equal to 188 4/7 reals vellon.
100 reals vellon ・・・・・・・・・・ 53⅛ do. plate.
17 reals plate ・・・・・・・・・・・・ 32 do. vellon.
17 piasters or current dollars 256 do. do.
4 maravadies make 1 quarto, 8½ quartos or 34 maravadies 1 real.
The peso, piaster, or current dollar of 8 reals plate, passes at 15 reals vellon in trade, but in exchange it is estimated at 15 reals vellon 2 maravadies.
The ducat of exchange is 375 maravadies.
The real plate, is estimated 10 cents, and the real vellon at 5 cents, in the United States.
The Spanish arobe, is 25 lb.
100 lb. of Spain is 97 lb. English.

・・・・・・

To change reals vellon to reals plate.

RULE. Multiply the given sum by 17, and divide by 32 for reals plate.

EXAMPLE.

Change 800 reals vellon to reals plate.

```
       800
        17
      ────
32)13600(425
   128
   ───
    80
    64
    ──
    160
    160
    ───
```
Ans. 425 reals plate.

To change reals plate to reals vellon.

RULE. Multiply the given sum by 32, and divide by 17 for reals vellon.

EXCHANGE.

EXAMPLE.

In 425 reals plate, how many reals vellon?

```
    425
     32
   ----
    850
   1275
   ----
17)13600(800
   136
   ----
    00
```

Ans. 800 reals vellon.

To change reals plate and reals vellon, to Federal money.

RULE. Multiply the reals plate by 10, and the reals vellon by 5, for the cents in the given sum.

EXAMPLES.

1. Change 14958 reals plate, to Federal money.

```
  14958
     10
  -----
  1495,80
```
Ans. 1495 dols. 80 cts.

2. Change 17593 reals vellon, to Federal money.

```
  17593
      5
  -----
  879,65
```
Ans. 879 dols. 65 cts.

......

CADIZ.

Accounts are kept by some in hard or plate dollars, reals vellon, and quartos.

8½ quartos ········ make ········ 1 real vellon.
20 reals vellon ················ 1 dollar of plate.

Others keep their accounts in reals plate and maravadies, reckoning 34 maravadies to 1 real plate.

To bring reals plate to dollars.

RULE. Multiply the given sum by 32, and divide by 17 for reals vellon, and divide the reals vellon by 20 for dollars.

EXCHANGE.

EXAMPLE.

In 320 reals plate how many hard dollars?

```
      320
       32
      ───
      640
      960
      ─────
17)10240(602 reals vellon
   102
   ───
    40
    34
    ──
     6
     8½          2|0)60|2 reals vellon
17)51(3 quartos.  dol. 30 2 3
   51
   ──
```

Ans. 30 dol. 2 r. v. 3 q.

To change hard dollars to reals plate.

Rule. Multiply the dollars by 20 for reals vellon, and the reals vellon being multiplied by 17 and divided by 32 give the reals plate required.—Or, Multiply the dollars by $10\frac{5}{8}$ for reals plate.

EXAMPLE.

In 16 hard dollars how many reals plate?

```
     16         Or thus,    16                16
     20                    10⅝                 5
    ───                    ───                ──
    320                    160               8)80
     17                     10                ──
    ───                    ───                10
   2240                   170 R. P.
    320
   ─────
32)5440(170
   32
   ───
    224
    224
```

Ans. 170 reals plate.

EXCHANGE. 181

Practical Questions,

The answers to which are in dollars, reals vellon, and quartos.

1. What will 45940 pipe staves come to at 80 piastres or current dollars per M. or 1200?

```
        45940
           80
12,00)367520,0
        3062⅔ current dollars.
           8   reals.
       24501⅓ reals plate.
          32
       49002
       73503
         10⅔
17)784042⅔(46120
    68
    ---
    104
    102
    ---
     20
     17
     ---
      34
      34
      ---
       2⅔
       8⅓
       ---
    17)22⅔(1
       17
       ---
        5⅔
```

```
2,0)4612,0
dols. 2306 0 1
```

Ans. 2306 h.dols. 0 r. 1 q.

		piast.			D.	R.	Q.
2.	21800 barrel staves at	30½	per 1200	....	417	3	7
3.	1200 hhd. do.	40	do.	....	30	2	3
4.	2 casks sherry wine	30	per cask	....	45	3	4

Q

182 EXCHANGE.

The result of the following is in reals plate, and maravadies.

5. In 610 hard dollars, how many reals plate?

$$610$$
$$20 \text{ reals vellon} = 1 \text{ hard dollar.}$$
$$\overline{12200}$$
$$17$$
$$\overline{85400}$$
$$12200$$
$$\overline{32)207400(6481}$$
$$192$$
$$\overline{154}$$
$$128$$
$$\overline{260}$$
$$256$$
$$\overline{40}$$
$$32$$
$$\overline{8}$$

Ans. 6481 r.p. 8 mar.

6. What will 2632 barrels of flour come to, at 11 current dollars per barrel?

$$2632$$
$$11$$
$$\overline{28952} \text{ piastres or current dollars.}$$
$$8 \text{ reals plate} = 1 \text{ piastre or current dol.}$$

Ans. 231616 reals plate.

7. 88 lasts of white dry salt, at 6 piastres per last.

$$88$$
$$6$$
$$\overline{528}$$
$$8$$
$$\overline{4224}$$

Ans. 4224 reals plate.

EXCHANGE.

8. Change £.600 sterling to reals plate, exchange at 36¼d. sterling per piastre.

```
              600
               20
            ─────
            12000
             - 12
            ─────
    36¼    144000
     4          4
    ─── ─────────
    145 ) 576000 ( 3972 current dollars.
          435        8
          ────     ────
          1410     31776
          1305       3 10
          ────     ──────
          1050     31779 10
          1015
          ────
           350
           290
          ────
            60
             8
          ────
    145)480 ( 3 reals.
        435
        ───
         45
         34
        ───
        180
        135
        ───
    145)1530(10 maravadies.
        145
        ───
         80        Ans. 31779 r.p. 10 mar.
```

9. In £.3200 sterling how many reals plate, exchange at 36¼d. sterling per piastre? Ans. 169489 r. p. 22 mar.

N. B. In St. Lucar accounts are kept in Reals plate and Quartos, 16 quartos to 1 real plate.

EXCHANGE.

BILBOA.

Accounts are kept in Reals vellon and Maravadies, 34 maravadies making 1 real.

The pound in Bilboa consists of 17 oz. except in iron which is but 16 oz.

    32 velts are equal to 66 gallons in the U. States.
    100 fanagues ...... 152 bushels   do.
    100 varas .... 108 yards   do.

To change piastres or current dollars to reals plate.

RULE. As 1 current dollar is to 15 reals 2 maravadies, so is the given sum to the reals required; or, multiply the sum by 15 reals 2 maravadies, for reals.

EXAMPLE.

In 5000 current dollars, how many reals vellon?

```
2 = 1/17 )5000              Or thus,    5000
       15  2 = 1 c. dol.                   2
       ─────                             ─────
       25000                      34)10000
        5000
         294  4                         294  4
       ─────                             
       75294  4                  Ans. 75294 r.vel. 4 mar.
```

To change current dollars to sterling.

RULE. As 1 dollar is to the rate of exchange, so is the given sum to the sterling required.

EXAMPLE.

In 5000 piastres or current dollars, how many pounds sterling, exchange at 36¾d. per dollar?

```
     p.      d.      p.
As   1   :  36¾  ::  5000
                      36¾
                    ──────
                    180000          5000
                      1875             3
                    ──────          ─────
                 12)181875        8)15000
                    ──────          ─────
                  2|0)15156  3'     1875
```

Ans. £.757 16 3.

EXCHANGE.

To change sterling to current dollars.

RULE. As the rate of exchange is to 1 dollar, so is the given sum to the dollars required.

EXAMPLE.

In £.757 16s. 3d. sterling, how many current dollars, exchange at 36⅜d. sterling per dollar?

```
    d.      dol.     £.  s.  d.
As 36¼  :   1   ::  757 16  3     Ans. 5000 cur. dols. or piast.
```

. . . . . .

To change sterling to reals vellon.

RULE. As the rate of exchange is to 15 reals 2 maravadies, so is the given sum to the reals required.

EXAMPLE.

In £.436 10s. sterling, how many reals vellon, exchange at 36⅜d. sterling per current dollar?

```
          d.      r. m.     £.   s.
    As   36¼  :  15  2  ::  436 10
          8                  20
         ───                 ───
         291.                8730
                              12
                            ──────
                            104760
                               8
2 marv. = 1/7               838080
                             15  2
                            ──────
                            4190400
                             838080
                              49298
               291)12620498(43369
                   1164
                   ────
                    980
                    873
                   ────
                   1074
                    873
                   ────
                   2019
                   1746
                   ────
                   2738
                   2619
                   ────
                    119
           34 mar. = 1 rial.
               294)4040(13.
```

Or, 838080
      2
   ──────
34)1676160 mar.
   ──────
    49298 reals.

Ans. 43369 reals 13 mar.

EXCHANGE.

PRACTICAL QUESTIONS.

1. What will 122 quintals of fish come to, at 136 reals per quintal?

```
    122
    136
   ----
    732
    366
    122
   -----
```
Ans. 16592 reals.

2. What is the cranage of 1137 quintals of fish, at 10 maravadies per quintal? Ans. 334 R. 14 M.

BARCELONA.

The monies of account in Barcelona and throughout the Province of Catalonia are Livres, Sols and Deniers.

12 deniers ......make...... 1 sol.
20 sols .... 1 livre.
37½ sols, or 1⅞ livre .... 1 hard dollar.
28 sols .... 1 cur. dol. the piast. of exchange.

To change livres to hard dollars.

RULE. Divide the livres by 3 and then by 5 and add the two quotients together for hard dollars.

EXAMPLES.

1. How many hard dollars in 360 livres?

```
3 | 360
  -----
5 | 120
  |  72
  -----
    192       Ans. 192 hard dols.
```

2. How many hard dollars must be paid for an invoice of goods amounting to 7134 livres?

```
3 | 7134
  ------
5 | 2378
  | 1426⅘
  ------
    3804⅘    Ans. 3804 h.d. 30 sols.
```

EXCHANGE.

To change hard dollars to livres.

RULE. Add to the given sum, the half, quarter, and eighth of it, and the sum will be the livres required.

EXAMPLES.

1. In 192 hard dollars, how many livres?

1/2	192
1/4	96
1/8	48
	24

   360.   Ans. 360 livres.

2. How many livres in 3804½ hard dollars?

1/2	3804,8
1/4	1902,4
1/8	951,2
	475,6

   7134,0.   Ans. 7134 livres.

To change livres to current dollars.

RULE. Multiply the livres by 5 and divide that product by 7 for current dollars.

EXAMPLE.

Change 2716 livres to current dollars.

$$\begin{array}{r} 2716 \\ 5 \\ \hline 7)13580 \\ \hline 1940 \end{array}$$

Ans. 1940 cur. dols.

To change current dollars to livres.

RULE. Multiply the current dollars by 7 and divide the product by 5 for livres.

EXAMPLE.

Change 1940 current dollars to livres.

$$\begin{array}{r} 1940 \\ 7 \\ \hline 5)13580 \\ \hline 2716 \end{array}$$

Ans. 2716 livres.

EXCHANGE.

PORTUGAL.

Accounts are kept in Millreas and Reas, reckoning 1000 reas to 1 millrea of 5s. 7½d. sterling, or 1 dol. 25 cts. in the U. States. A vinten is 20 reas, and 5 vintens is a festoon of 100 reas.

---

1. Change 579 millreas 740 reas to Federal, at 1 dol. 25 cts. per millrea.

           M. R.
   579,740 Or thus, 579,740
     1,25   ¼ added 144,935
   ―――――   ―――――
   2898 700   Dollars 724,675
   69568 80
   ―――――
Cents 72467,500   Ans. 724 dols. 67½ cts.

2. Change 724 dols. 67½ cts. to millreas, at 1 dol. 25 cts. per millrea.

    1,25)724,675(579 mill. 740 reas.

Or, deducting ⅕ from the sum in Federal money gives the millreas, &c.

  Example. ⅕)724,675
        144,935
       ―――――
       579,740 as before.

3. Change 579 millreas 750 reas to sterling, at 5s. 7½d. per millrea.

      579,750
        67½
     ―――――
     4058,250
     34785,00
      289,875
    ―――――
   12)39133,125

   2|0)3261|1
   ―――――
  Ans. £.163 1 1⅛

4. In £.163 1 1⅛ sterling, how many millreas, at 5s. 7½d. per millrea?

   s. d. reas. £. s. d.
   5 7½ : 1000 :: 163 1 1⅛
        Ans. 579 mill. 750 reas.

EXCHANGE.

5. What is the commission on 6245 mill. 46 reas, at $2\frac{1}{2}$ per cent. ?

$$6245,046$$
$$2\frac{1}{2} \text{ per } 1,00$$
———
12490092
3122523
———
156,12615   Ans. 156 mill. 126 reas.

6. Suppose a cargo is sold for 6245 millreas, at 2 months credit, for prompt payment of which $\frac{1}{2}$ per cent. per month is allowed; how much is the discount?

$\frac{1}{2}$)6245   Or thus,
——— $\frac{1}{2}$ per cent. for 2 months = 1 per cent.
31,225 for 1 month.   6245
   2   1
———   ———
Ans. 62,450 for 2 months.   62,45

7. Suppose you import 5960 hhd. staves and 5060 barrel staves on which there is a duty of 23 per cent. which is taken in kind, how many of each remain for sale?

Ans. 4590 hhd. and 3897 bbl.

		M. R.		M. R.
8.	702 barrels of flour at	8,600 per bbl.	....	6037,200
9.	4590 hhd. staves	,030 per stave	....	137,700
10.	3897 bbl. do.	,020 per do.	......	77,940
11.	71 alquiers of beans ..	,480 per alquier	..	34,080

MEASURES OF PORTUGAL.

Cloth Measure.

A vara is $43\frac{1}{9}$ inches English.
A covedo is $20\frac{2}{3}$ ditto.

Wine Measure.

1 almude is 12 canados.
1 canado is 4 quarteels.
An almude is $4\frac{1}{2}$ gallons English wine measure.
A canado is 3 pints English.

Corn Measure.

1 moy is 15 fangas.
1 fanga is 4 alquiers.
1 moy of 60 alquiers is 3 English quarters, or 24 bushels Winchester measure.
1 quarter is 20 alquiers.
1 English bushel is 2½ alquiers in Lisbon, 2 alquiers in Oporto, and 2⅔ alquiers in Figuiras.
A moy of salt is the same measure as corn.
A pipe of coals is 16 fangas.
1 fanga is 8 alquiers.
A pipe of coals is 128 alquiers, which at 2½ alquiers per bushel, is 51½ bushels English.

WEIGHTS OF PORTUGAL.

1 quintal is 4 arobes.
1 arobe is 32 pounds, so that a quintal is 128 lb. Portugal wt. which is equal to about 132 lb. English, avoirdupois weight.
A pound is about 16½ ounces English.

Loss by exchanging English money in Portugal.

An English guinea passes at Lisbon for 3 m. 600 r. which is 134 reas, or 9 pence less than the value.
An English crown passes for 800 reas, which is 89 reas, or 6 pence less than the value.
An English shilling passes for 160 reas, which is 18 reas, or about 1¼ penny less than the value.

* * * * *

LEGHORN.

Accounts are kept in Piastres, Soldi, and Denari, reckoning 12 deniers to 1 soldi, and 20 soldi to 1 piastre or dollar of 48d. sterling at par.

1½ paul, or 2 sols, are equal to 1 livre.
6 livres .... 1 piastre or dollar.
5¾ livres (effective money) .... 1 do.
1 ducat .... 1¾ do.

EXCHANGE.

Weights—A pound is only 12 ounces in all commodities.
145 lb. is said to be equal to the English quintal of 112 lb.; but fish generally renders about 136 to 138 lb. per quintal.

145 lb. in Leghorn   make   112 lb. in the U. States.

4 brasses .... 1 cane.
100 brasses .... 64 yards, U. States.
1 palm .... 9½ inches, do.

4 sacks are 2 per cent. less than an English quarter, of 8 bushels.

---

1. How much will 5630 lb. of ginger come to, at 9 piastres per 100?

$$\begin{array}{r}5630\\9\\\hline 506|70\\20\\\hline 14|00\end{array}$$

Ans. 506 piast. 14 sol.

2. What will 9760 lb. of pepper come to, at 27¼ ducats per 100?

$$\begin{array}{r}9760\\27\tfrac{1}{4}\\\hline 68320\\19520\\2440\\\hline \tfrac{1}{6})265960\\44326\tfrac{2}{3}\end{array}$$

piast. 3102|86⅔
20
─────
soldi 17|33⅓
12
─────
den. 4|00   Ans. 3102 piast. 17 sol. 4 den.

EXCHANGE.

3. What will 143700 lb. of pitch come to, at 26 pauls per 100?

NOTE. 1 paul is equal to ⅔ of a livre.

```
    143700
        26
    ──────
    862200
   287400
   ───────
  37362,00 pauls.
        2
   ───────
 3)74724
   ──────
 6)24908 livres.
   ──────
   4151 6 8
```

Ans. 4151 piast. 6 sol. 8 den.

4. How much will 4200 sacks of wheat come to, at 26 livres, effective money, per sack?

```
    4200
      26
   ─────
   25200
    8400
```

liv.   piast.
5¾ :   1   ::   109200 livres.

Ans. 18991 piast. 6 sol. 1 den.

				piast.	s.	d.
5.	100 barrels pork	16 piastres per barrel		1600	0	0
6.	1000 do. flour	10½ do.		10500	0	0
7.	2660 lb. coffee	26 do. per 100		691	12	0
8.	6578 lb. pimento	18 do. do.		1184	0	9
9.	9370 lb. rice	24 liv. cur. money per 100		374	16	0
10.	97270 lb. logwood	16 piastres per 1000		1556	6	4
11.	4170 lb Russia wax	33½ ducats per 100		1629	15	6
12.	104060 lb. sugar	30 piastres per 151 lb.		20674	3	5
13.	3350 lb. loaf sugar	30 do. per 100		1005	0	0
14.	1000 casks tar	4½ do. per cask		4500	0	0
15.	100000 staves	4 do. per 100		4000	0	0

EXCHANGE.

NAPLES.

Accounts are kept in Ducats and Grains, reckoning 100 grains to 1 ducat.

The current coins are grains, carlins, ducats, dollars, and ounces.

10 grains make 1 carlin; 10 carlins 1 ducat; 3 ducats 1 ounce.

The Naples dollar passes for 120 grains, and the Spanish dollar for 126 grains.

100 lb. Naples weight are equal to $64\frac{5}{8}$ lb. English.

Brandy is sold per cask of 12 barrels, or 132 gallons; 60 karafts make a barrel.

Sewing silks are sold per lb. of 12 ounces.

Lustrings are sold per cane of 84 inches.

Sugar, coffee, fish, and tobacco, are sold per cantar, of 196 lb. in the United States.

The cantar is subdivided into 100 rotolas of 33 ounces each.

---

1. What is the amount of 10 casks 6 barrels 29 karafts of brandy, at 92 ducats per cask?

$$\begin{array}{r} 92 \\ 10 \\ \hline 920 \end{array}$$

6 bbl. $\frac{1}{2}$ 46
20 kar. $\frac{1}{18}$ 2 55
5 do. $\frac{1}{4}$ 64 nearest.
4 do. $\frac{1}{5}$ 51
             ―――
            969 70   Ans. 969 ducats, 70 grains.

2. What is the amount of 2 casks of clayed sugar, weighing neat 10 cantars 51 rotolas, at 65 dollars per cantar?

rot.   dols.   rot.
100 : 65 :: 1051   Or thus,   65
              65   10
            ―――   ―――
           5255   650
           6306   50 rot. $\frac{1}{2}$ 32 50
           ―――   1 do. $\frac{1}{50}$ 65
   duc. 683,15   ―――
                            duc. 683 15

Ans. 683 ducats, 15 grains.

R

EXCHANGE.

3. How much is the amount of 1 box of scented soap, containing 100 parcels of 16 ounces each, at 22 grains per rotola?

$$100$$
$$16$$

oz. gr.
33 : 22 :: 1600 oz. : Ans. 10 ducats, 66 grains.

4. What is the commission on 996 ducats, at 2 per cent.?
Ans. 19 ducats, 92 grains.

	can. rot.		ducats.	due.	gr.
5.	3 73	of coffee	73 per cantar ..	272	29
6.	16 19¾	soap	21	340	14
7.	1 59	do..............	21	33	39
8.	7 97¾	do..............	21	167	52
9.	67½	scented ditto	30	20	25
10.	52	white ditto	17	8	84
11.	7 64	raisins	12	91	68
12.	2 casks 11 bbls. 4 kar. of brandy	102 per cask ..	298	06
13.	......10 do. 43 do. ditto ..	92 do.	82	16
14.	...... 9 do. 12 do. ditto ..	92 do.	70	53
15.	355 canes of silk	2 50 per cane	887	50

* * * * *

TRIESTE.

Accounts are kept in Florins and Kreutzers—60 kreutzers make 1 florin.

The exchange on London, (8th July, 1803,) was 12 florins for the pound sterling.

The other kinds of money are Soldi and Livres.
20 soldi .... make .... 1 livre.
5¼ livres .... 1 florin.

100 lb. Vienna weight = 123 lb. Avoirdupois.
A brace is 27 inches, or ¾ of a yard English.
A barrel of wine is 18 gallons.
A staro of wheat is 2⅜ bushels nearly—3⅓ staros is equal to an English quarter of 8 bushels.

Sales and purchases are usually made in bills on Vienna at *3 months* date.

EXCHANGE.

1. What is the amount of 263 lb. Vienna weight, of soap, at 22 kreutzers per lb. ?

$$\begin{array}{r} 263 \\ 22 \\ \hline 526 \\ 526 \\ \hline 6|0)578|6 \\ \hline 96\ 26 \end{array}$$

Ans. 96 flor. 26 kreutzers.

2. 758 gallons wine, at 21 florins 30 kreutzers per barrel ?

$$\begin{array}{r} 758 \\ 21 \\ \hline 758 \\ 1516 \\ 30\ \text{kr.} \quad \tfrac{1}{2} \quad 379 \\ \hline 18)16297(905 \\ 162 \\ \hline 97 \\ 90 \\ \hline 7 \\ 60 \\ \hline 18)420(23 \\ 36 \\ \hline 60 \\ 54 \\ \hline 6 \end{array}$$

Ans. 905 fl. 23⅓ kr.

                       fl. kr.   fl. kr.

3. 120 staros of wheat at 4 20 per staro. Ans. 520 00
4. 715 braces of silk ···· 3 50 per brace. ···· 2740 50
5. 1730 lb. coffee ······   58 per lb. ······ 1672 20

GENOA.

Accounts are kept in **Denarii**, **Soldi**, and **Pezzos** or **Lires**.
 12 denarii .... make .... 1 soldi.
 20 soldi .... 1 pezzo or lire.
 1 pezzo of exchange .... $5\frac{3}{4}$ lires.

The course of exchange is various—from $47d.$ to $58d.$ sterling per pezzo or lire.

In Milan,	1 crown	=	80 soldi of Genoa.
.. Naples,	1 ducat	=	86 do.
.. Leghorn,	1 piastre	=	20 do.
.. Sicily,	1 crown	=	$127\frac{2}{3}$ do.

To reduce Exchange money to Lire money.

RULE. Multiply the exchange money by $5\frac{3}{4}$ for lire money.

EXAMPLE.

In 384 pezzos of exchange how many lires?

```
      384
       5¾
     ----
     1920
¾    192
¼     96
     ----
     2208        Ans. 2208 lires.
```

To reduce Lire money to Exchange.

RULE. Multiply the lire money by 4 and divide the product by 23 for exchange.

EXAMPLE.

In 2208 lires how many pezzos of exchange?

```
     2208
        4
    -----
 23)8832(384
    69
    ---
    193
    184
    ---
     92
     92
     --     Ans. 384 pezzos of exchange.
```

EXCHANGE.

To reduce Lires to Sterling.

RULE. As 1 lire is to the rate of exchange so is the lires to the sterling required.

EXAMPLE.

In 360 lires how much sterling, exchange at 54d. sterling per lire?

```
    l.      d.        l.
    1   :   54   ::   360
                       54
                     ─────
                     1440
                     1800
                    ──────
                 12)19440
                    ──────
                 2|0)162|0
                    ──────
                    £.81        Ans. £.81 0 0 sterl.
```

VENICE.

Venice has three kinds of money, viz. Banco money, Banco current money, and Picoli money. Banco money is 20 per ct. better than banco current, and banco current 20 per ct. better than picoli.

The different denominations of money are Denari, Soldi, Grosi, and Ducats.

12 denari, or deniers d'or, make 1 soldi, or sol d'or.
$5\frac{1}{0}$ soldi .... 1 gros, or grosi.
24 gros, or grósi .... 1 ducat.

100 ducats banco of Venice in Leghorn = 93 pezzos.
.... Rome = $68\frac{1}{2}$ crowns.
.... Lucca = 77 do.
.... Frankfort = $139\frac{1}{2}$ florins.

The par of exchange in 1798 was $50\frac{1}{4}d$. sterling per ducat banco.

EXAMPLE.

How much sterling is equal to 2712 ducats banco, exchange at 50¼d. sterling per ducat banco?

```
 duc.       d.       duc.
  1  :    50¼   ::   2712
           4          201
         ─────       ─────
          201        2712
                    54240
                   ───────
              4)545112 farth.
              ──────────────
             12)136278 pence.
             ───────────────
             2|0)11351|6  6 shills.
             ─────────────
       Ans. £.567  16  6 sterling.
```

. . . . .

SMYRNA.

Accounts are kept in piastres and hundredths, except the English accounts, which from ancient custom are kept in piastres and eightieths or half paras.

The fractional parts are sometimes called aspers, 100 aspers to 1 piastre.

The following calculations are made in piastres and hundredths.

A piastre is equal to 40 paras, and a Spanish dollar to 136 paras.

340 piastres are equal to 100 Spanish dollars.

The exchange on London was 13 piastres for 1 pound sterling, May 14th, 1800.

Their weights are the Rotola, Oke, Cheque and Tiffec—

A rotola .... marked *Ro.* is 180 drams.
An oke .... % is 400 do.
A cheque of opium .... is 250 do.
   do. of groat's wool............. is 800 do. or 2 okes.
A tiffee of silk .... is 610 do.

100 rotolas, or 18000 drams, or 45 okes are a quintal of this country.

112 lb. English should render here 40¾ okes, or 90⅖ rotolas.

45 okes of this country render 123¾ lb. English.

A *pike* is 27 inches nearly.

EXCHANGE.

To change piastres to dollars.

RULE. Multiply the piastres by 5, and divide the product by 17, for cents.

EXAMPLE.

Change 1277 $\frac{55}{100}$ piastres to dollars.

```
    1277,55
          5
    ─────────
17)6387,75(375,75
   51
   ──
   128
   119
   ───
    97
    85
    ──
    127
    119
    ───
     85
     85
```
Ans. 375 dols. 75 cts.

To change dollars to piastres.

RULE. Multiply the dollars by $3\frac{2}{5}$ for piastres.

EXAMPLE.

Change 375 dollars 75 cents to piastres.

```
   375,75
      3⅖
   ──────
   1127,25
     75,15 ⎫
     75,15 ⎬ for ⅖
   ──────
```
Ans. 1277,55 piastres.

· · · · · ·

PRACTICAL QUESTIONS.

1. How much will 10 serons of cochineal come to, weighing neat 724 okes 73 rotolas, at 80 piastres per oke?

```
   724,73
       80
   ──────
```
Ans. 57978,40 piastres.

EXCHANGE.

2. 299 bags of sugar, weighing 506 quintals 96 rotolas, tare 14 rotolas per bag, at 110 piastres per quintal.

```
      gross  506 96              299
      tare    41 86               14
                                 ----
      neat   465 10             1196
              110                299
             ------             ----
  Ans. 5116L 00 piast.       100)4186
                                  ----
                                  41 86
```

3. 4 cases of opium, weighing gross 1026 rotolas, tare 84 okes 75 rotolas, at 10¾ piastres per cheque.

Note. 1 rotola is equal to $\frac{9}{20}$ of an oke, and 1 oke to 1¾ cheque.

```
        rot.   1026
                  9
              ------
           20)9234 rot.
              ------
  gross okes 461 70
  tare        84 75
              ------
  neat okes  376 95         376 95
              1¾              3
              ------         ------
             376 95        5)1130 85
             226 17           ------
             ------           226 17
  cheques    603 12
              10¾
             ------
            6031 20
             301 56
             150 78
```

Ans. piast. 6483 54.

4. 893 pieces of copper, neat okes 19743,85, at $\frac{70}{40}$ or 70 paras per oke.   O. R.

```
              19743,85
                    70
              --------
           4|0)1382069|5|0
```

Ans. piast. 34551,73

EXCHANGE. 201

5. What is the custom-house duty on 19740 okes of copper at $\frac{2\frac{1}{2}}{40}$ agio $2\frac{1}{2}$ per cent. ?

NOTE. The charges are all established by a tariff of the Levant Company.

$$19740$$
$$2\frac{1}{2}$$
$$\overline{39480}$$
$$9870$$
$$\overline{4|0)4935|0}$$

agio $2\frac{1}{2} = \frac{1}{40}$) 1233,75 amount of duty at $2\frac{1}{2}$ paras.
30,84 agio at $2\frac{1}{2}$ per cent.

Ans. piast. 1264,59

6. English consulage on 430 quintals, at $5\frac{1}{2}$ piast. agio 7 per cent.

$$430$$
$$5\frac{1}{2}$$
$$\overline{2150}$$
$$215$$
$$\overline{2365}$$
$$7$$

Ans. piast. 165,55

7. Custom-house duties on 88 quintals 90 rotolas, at $\frac{20}{110}$, agio $2\frac{1}{2}$ per cent.

$$88,90$$
$$20$$
$$\overline{11|0)17780|0}$$
$$2\frac{1}{2} = \frac{1}{40})16,16$$
$$,40$$

Ans. piast. 16,56

8. What will the following charges amount to, viz. porterage $\frac{8}{40}$, house porters $\frac{4}{40}$, weighing $\frac{2}{40}$, chan duty $\frac{2}{40}$, visiting and marketing $\frac{1}{40}$ per quintal on 438 quintals?

```
porterage .... 8              438
house porters  4               17
weighing ....  2              ----
chan duty....  2           4|0)744|6
visiting ....  1              ----
              --
              17         Ans. piast. 186,15.
```

ENGLISH WEST-INDIES.

Accounts are kept in Pounds, Shillings, and Pence.

JAMAICA AND BERMUDAS.

The Spanish dollar passes at 6s. 8d.; 3 dollars are equal to 20 shillings, or 1 pound, Jamaica currency.

To change Jamaica currency to Federal.

RULE. Multiply the pounds by 3 for dollars. If there be shillings, &c. increase the pence in the given sum by $\frac{1}{4}$ for cents.

EXAMPLES.

1. When lumber is sold in Jamaica at £.15 per M. how much is it in Federal money?

```
         15
          3
         --
Ans.  45 dols.
```

2. Change £.54 12s. 11d. Jamaica currency to Federal.

```
    54 12 11
    20
    ------
    1092
    12
    ------
 ¼)13115
    3278¾
    ------
   16,393¾ cts.      Ans. 163 dols. 93¾ cts.
```

EXCHANGE. 203

3. What will 102,896 feet of boards come to, at £.15 per M.?

$$\begin{array}{r}102,896\\15\\\hline 514480\\102896\\\hline £.1543,440\\20\\\hline s.\ 8,800\\12\\\hline d.\ 9,600\end{array}$$

Ans. £.1543 8 9

4. What will 5 hhds. of sugar come to, weighing 8519 lb. neat, at 70 shillings per 100 lb.?

$$\begin{array}{r}8519\\70\\\hline 2|0)5961|3,30\\\hline\end{array}$$

Ans. £.298 3 3

5. How much will 5 hhds. of sugar come to, weighing 9103 lb. neat, at 75 shillings per 100 lb.?

$$\begin{array}{r}9103\\75\\\hline 45515\\63721\\\hline 2|0)6821|7,25\\\hline\end{array}$$

Ans. £.341 7 3

. . . . .

BARBADOES.

The Spanish dollar is 6s. 3d. Barbadoes currency.

EXCHANGE.

To change Barbadoes currency to Federal.

RULE. Increase the pence in the given sum by ⅓ for cents.

EXAMPLE.

Change £.49 11s. 10d. Barbadoes money to Federal.

£.49 11 10
20
―――
991
12
―――
⅓)11902
3967⅓
―――
158,69⅓

Proof ⅓)15869⅓ cents.
3967⅓
―――
12)11902 pence
―――
2|0)991|1 10
―――
£.49 11 10

Ans. 158 dols. 69⅓ cents.

Other calculations as in Jamaica.

MARTINICO, TOBAGO, AND ST. CHRISTOPHERS.

These islands being inhabited by French and English, the former keep their accounts in Livres, Sols, and Deniers, and the latter in Pounds, Shillings, and Pence.

A *current* dollar   is   8s. 3d.
A *round* dollar   passes for   9s.

When payment of freight or goods is mentioned in Spanish dollars, disagreement respecting their value has frequently arisen; and to prevent it, some persons distinguish them by *round* and *current* dollars; others mention the *bits* to each. But the most certain way is to specify the number of shillings or livres, instead of dollars; thus A sells to B a barrel of flour, at 99 shillings or livres; in payment B may allow him 11 dollars at 9 shillings each, or 12 dollars at 8s. 3d. each, either being equal to 99 shillings or livres, the sum specified by their agreement.

FRENCH WEST-INDIES.

Accounts are kept in Livres, Sols, and Deniers.
12 deniers make 1 sol, and 20 sols 1 livre.
The Spanish dollar passes in some places for 8 livres 5 sols, and in others for 9 livres.

EXCHANGE.

1 cwt. or 112 lb. in the U. States is equal to 104 lb. French. 100 lb. French are equal to 108 lb. nearly, in the U. States.

When any commodity is to be marked in French weight, 4 per cent. is added to the neat hundreds; thus a hogshead of fish weighing neat 10 cwt. is marked 1040 lb. Fish shipped from the United States will answer to the weight thus marked, provided it comes out in good order, and the *cask* weighs exactly the customary tare, which is 10 per cent.

100 lb. of coffee or cotton, bought in the French islands, will, or ought to weigh 108 lb. (it will often weigh 110 lb.) in the United States; and as these articles are sold here per lb. there is a gain of 8 to 10 per cent. in the weight. But on sugar, which is bought for 100 lb. and sold here per 112, there is a loss of 6 per cent. because there is 4 per cent. between the American cwt. and 100 lb. French, and 2 per cent. difference in the tare. The tare on brown sugar in the French islands being 10 per cent. and the American tare 12 per cwt. The loss on clayed sugar is greater, occasioned by the customary tare, which is but 7 per cent. in the French islands, whereas it is here 12 per cent. the same as on brown sugar.

NOTE. The tare allowed on sugar among merchants is 12 per 112; that allowed by the custom-house is 12 per 100. {*See Tare and Tret*, page 95.}

1. Change 10692 livres to dollars, at $8\frac{1}{4}$ livres per dollar.

$$
\begin{array}{r}
8\frac{1}{4} \quad 10692 \\
4 \quad\quad 4 \\
\hline
33 \,)\, 42768\,(1296 \\
33 \\
\hline
97 \\
66 \\
\hline
316 \\
297 \\
\hline
198 \\
198 \\
\hline
\end{array}
$$

Ans. 1296 dols.

S

EXCHANGE.

2. Change 7713 livres to dollars, at 9 livres per dollar.

$$9)7713$$

Ans. 857 dollars.

3. In 1296 dollars, at 8¼ livres each, how many livres?

```
   1296
     8¼
  ─────
  10368
    324
  ─────
```
Ans. 10692 livres.

4. 857 dollars, at 9 livres each, how many livres?

```
  857
    9
 ────
```
Ans. 7713 livres.

5. What will 1642 lb. of coffee come to at 15 sols per lb.?

```
  1642
    15
 ─────
  8210
  1642
 ─────
 2|0)2463|0 sols.
```
livres 1231 10   Ans. 1231 liv. 10 sols.

6. 1780 lb. cotton at 157 livres 10 sols per 100 lb.

```
          1780
           157
         ─────
         12460
          8900
          1780
10 sols. ½   890
         ─────
   liv.  2803|50
             20
         ─────
   sols   10|00
```
Ans. 2803 liv. 10 sols.

EXCHANGE.

7. 24 barrels of beef at 101 liv. 1 sol 3 den. per barrel.

```
    liv.   s.   d.
    101    1    3
                6
    ─────────────
    606    7    6
                4
    ─────────────
    2425  10    0     Ans. 2425 liv. 10 sols.
```

8. How many dollars, at 8 livres 5 sols per dol. will pay for 12 hhd. of brown sugar, weighing 13365 lb. at 40 liv. per 100 lb.?

```
            13365
               40
            ─────
    8¼    534600,00
    4         4
    ─────────────
    33) 213484 (648 dols.
        198
        ─────
        158
        132
        ─────
         264
         264           Ans. 648 dols.
```

9. A cargo, amounting to 12536 dols. in the United States is sold at 12½ per cent. advance on the invoice; how many livres will it amount to, estimating the dollar at 8¼ livres each?

```
    12½=⅛) 12536  invoice.
            1567  advance.
           ─────
           14103  amount.
               8
           ──────
          112824  livres at 8 per dollar.
    5 sols  ¼  3525¾
           ──────
    Ans. 116349¾ livres at 8¼ per dollar.
```

		sols.	d.		liv	s.	d.
10.	6 hhds. coffee, weighing 4471 lb. at 14	6 per lb.			3241	9	6
11.	14 do. sugar, do. . . . . 16477 . . . . 38 liv. per 100				6261	5	2
12.	1 bale of cotton, do. . . 227 . . . . 150 . . . . do. . .				340	10	0
13.	94 hhds. fish, . . do. 101313 . . . . 33 . . . . do. . .				33433	5	9
14.	16 casks of rice, do. . . 6575 . . . . 40 10 . . do. . .				2662	17	6
15.	1390 hoops . . . . 480 . . . . per M.				667	4	0
16.	15059 feet of boards . . . . 100 . . . . do. . .				1505	18	0
17.	48 shaken hhd. with heads. . . . . 7 15 per hhd.				372	0	0
18.	29 barrels of beef . . . . 90 15 per bbl.				2631	15	0
19.	6759 velts of molasses . . . . 26 per velt				8786	14	0
20.	32670 gals. do. at 73l. 7s. 9d. per tierce of 60 gals.				39959	9	10

SPANISH WEST-INDIES.

Accounts are kept in Havanna, Laguira, Vera Cruz, &c. in dollars and reals, reckoning 8 reals to a dollar.

The Spanish arobe is 25 lb.

1. What will 123 pieces Bretagnes come to, at 26 reals per piece?

```
      123
       26
     ----
      738
     246
     ----
   8)3198
     ----
     399 6    Ans. 399 dols. 6 reals.
```

2. 21784 feet boards, at 45 dollars per thousand.

```
      21784
         45 per M.
     ------
     108920
      87136
     ------
     980|280
        8
     ------
      2|240.           Ans. 980 dols. 2 reals.
```

EXCHANGE.

3. 153 cases of gin, at $8\frac{6}{8}$ dollars per case.

$$\begin{array}{r} 153 \\ 8\frac{6}{8} \\ \hline 1224 \end{array}$$

4 reals   76   4
2 do.   38   2

1338   6.   Ans. 1338 dols. 6 reals.

4. What is the commission on 14792 dollars 5 reals, at 4 per. cent.?

14792   5
      4

591|70   4.
     8

5|64.   Ans. 591 dols. 5 reals.

5. What will 42 bbls. of white sugar come to, weighing gross 415 arobes 18 lb. tare and tret on the whole 858 lb. at 26 reals per arobe?

                ar.   lb.
                415   18
858 lb. make   34   8

                381   10
                 26

                2286
                 762

10 lb. $=\frac{2}{5}$ arobe   10

             8)9916 reals.

             1239   4.   Ans. 1239 dols. 4 reals.

                                          dols. reals.
6.   125 pieces bretagnes at 26 reals ············ 406   2
7.   500 do. ·· do. ···· $24\frac{1}{2}$ do············ 1531   2
8.   80 umbrellas ······ $6\frac{1}{2}$ dollars ······· 520   0
9.   147 arobes of butter··· 25   do. per 100 lb. 918   6
10.   2405 arobes 19lb. sugar 25 reals per arobe 7518   0
11.   1660 do. ·· 12 ·· do. ··· 21   do. ·· do. ·· 4358   7
12.   16695 feet boards ···· 40 dols. per M. ·· 667   6

S 2.

EXCHANGE.

EAST-INDIES.

CALCUTTA.

Accounts are kept in Rupees, Annas, and Pice.

12 pice make 1 anna, 16 annas 1 rupee.

By the *bazar*, or market exchange, for June, 1797, the exchange was, viz.—
100 English guineas were equal to 956 rupees 4 annas.
100 Spanish dollars were equal to 212 rupees.

In Weights.—16 chittacks make 1 seer, 40 seers 1 maud.

The factory maud is 74 lb. English.
The bazar maud is 82 ditto.
The imports are sold by the factory maud and current rupees.
The exports are bought by the bazar maud and sicca rupees.
100 sicca rupees are equal to 116 current rupees.

Bednah, tin-plates, and hides, are sold per corge, 20 to a corge.
The cavid is half a yard English.

. . . .

1. What will 3905 dry hides amount to, at 12 rupees per corge?

```
  h.   r.     h.
  20 : 12 :: 3905
              12
            ------
        2|0)4686|0
            ------
             2343        Ans. 2343 rupees.
```

2. How much will 189 bazar mauds 31 seers 8 chittacks of sugar come to, at 6 rupees per maud?

```
              189  31  8
                6
             --------
             1134
20 seers  ½     3
10        ¼     1  8
 1       1/20   0  2  4
 8 chit.  ½     0  1  2
             --------
             1138 11  6     Ans. 1138 r. 11 a. 6 p.
```

BOMBAY.

Accounts are kept in Rupees, Quarters, and Rees.

100 rees make 1 quarter; 4 quarters 1 rupee.

218 rupees were equal to 100 Spanish dollars, in April, 1800.

The current money is in Mohurs, Rupees, and Pice.

50 pice make 1 rupee; 15 rupees 1 mohur.

The weights are pounds, mauds, and candies; the pound the same as English.

A Bombay maud is 28 lb.

A Surat maud is $37\frac{1}{3}$ lb.

21 Surat mauds or 784 lb. make 1 Surat candy.

Cotton is sold by the Surat candy.

Camphire and Mocha coffee are sold by the Surat maud.

Malabar pepper is sold by the Bombay candy of 588 lb.

....

In 274 bales of cotton, weighing neat 996 cwt. 2 qrs. 23 lb. how many Surat candies?

784 lb. = 7 cwt.   7)996 2 23

      142 200 two hundreds.
          24 excess 12 per cent.
          56 two quarters.
          23
         ———
          303 Ans. 142 can. 303 lb.

MADRAS.

Accounts are kept in Pagodas, Fanams, and Cash.

80 cash make 1 fanam; 36 fanams 1 pagoda.

The Spanish dollars were in 1798 and '99, at 165 dollars for 100 star pagodas; making the pagoda worth 165 cents. The revenue laws of the United States reckon them at 184 cents.

The Bengal, or Sicca (new) rupee is worth 46 to 47 cents. The revenue laws of the United States value them at 50 cents.

EXCHANGE.

The current exchange is 340 Sicca rupees, for 100 Star pagodas.

A Lack of rupees is 100,000.

Cowries are sea shells used as small money in India, and on the coast of Africa, to make change among the natives in the bazar, or market, and in payment to the coolies or labourers. In May, 1792, a rupee was worth 5120 cowries. The common cowries are generally at 5 to 7 rupees per Bazar maud, the better sort from 10 to 14 rupees per maud, the price varying according to the kind.

The picul is 133⅓ lb. English.

100 cattas make a picul.

A maud is 25 lb. Troy, 20 mauds make 1 candy.

The excellence of their cloth is defined by the *threads* in the warp.

The duty payable at the custom-house is 2½ per cent. outwards and inwards. This is taken on imports according to the invoice, and on exports at the actual cost at the bazar or market.

* * * * * *

BATAVIA.

Accounts are kept in Rix Dollars and Stivers.

The rix dollar   is 48 stivers.
The ducatoon   is 80 ditto.
The Spanish dollar is 64 ditto; sometimes it passes at 60 stiv.

    125 lb. Dutch are equal to 133⅓ lb. English.
    125 do.   make   1 picul.
    100 cattas • • • • • • • • • • • • • 1 ditto.

* * * *

1. In 1333 rix dols. 16 stivers, how many ducatoons?

$$\begin{array}{r} 1333 \quad 16 \\ 48 \\ \hline 10670 \\ 5333 \\ \hline 8|C\,)6400|0 \\ \hline \end{array}$$

Ans. 800 ducatoons.

EXCHANGE.

2. What will 127477 cattas of bar iron come to, at 9 rix dollars per picul?

```
     cat.   r.d.      cat.
As   100  :  9  ::  127477
                         9
                   ───────
                   11472,93
                      48
                   ───────
                      744
                      372
                   ───────
                      44,64    Ans. 11472 r. dols. 44 st.
```

3. What will 3894 bottles of wine come to, at 36 stivers per bottle?

```
                3894                  Or thus, 36 stiv. = ⅔ rix dol.
              ──────                           3894
24 stiv.  ½   1947                                3
12        ¼    973  24                       ──────
              ──────                         4)11682
              2920  24                       ──────
                                             2920  24
Ans. 2920 rix dols. 24 stivers.
```

4. In 31478 lb. of sugar, how many piculs?

```
125)31478(251
    250
    ───
    647
    625
    ───
    228
    125
    ───
    103         Ans. 251 piculs 103 lb.
```

                                               pic. lb.
5. In 50632 lb. how many piculs?   Ans. 405   7
6.   12648 .... 101 23

7. What will 279 piculs 25 lb. of sugar come to, at 7½ rix dollars per picul?

```
              279
               7½
            ─────
             1953
              139 24
  25 = ⅓       1 24
            ─────
             2094 00        Ans. 2094 rix dols.
```

......

C H I N A.

Calculations are made in Tales, Mace, Candareens, and Cash.

10 cash ...... make ...... 1 candareen.
10 candareens .... 1 mace.
10 mace .... 1 tale.

The tale of China is estimated at 1 dollar 48 cents in the United States.

The Spanish dollar is current at 72 candareens.

Weights are in Tales, Piculs, and Cattas—
16 tales make 1 catta; 100 cattas 1 picul.

A picul is equal to 133⅓ lb. English.

The cavid of China is 14 $\frac{2}{10}$ inches; it is divided into 10 parts.

....

To change pounds English to Cattas.

RULE. Deduct 25 per cent. or one quarter, for cattas.

EXAMPLE.

In 62668 lb. English, how many cattas?

```
    ¼)62668
      15667
    ──────
```

Ans. 47001 cattas.

....

To change cattas to pounds English.

RULE. Add one third for pounds English.

EXAMPLE.

In 47001 cattas, how many lb. English?

```
    ⅓)47001
      15667
    ──────
```

Ans 62668 lb. English.

EXCHANGE.

Practical Questions.

1. What is the amount of 308 chests of bohea tea, weighing neat 101956 lb. at 15 tales per picul?

$$\tfrac{4}{3})101956 \text{ lb.}$$
$$25489$$

cat.	tal.		
100	: 15	::	76467 cattas.

$$15$$

$$382335$$
$$76467$$

11470,05   Ans. 11470 tales 5 cand.

2. What will 75 chests of souchong tea come to, weighing neat 4875 lb. at 44 tales per picul?

$$\tfrac{4}{3})4875$$
$$1218\tfrac{3}{4}$$

$$3656\tfrac{1}{4} \text{ cattas.}$$
$$44$$

$$14624$$
$$14624$$
$$11$$

1608,75   Ans. 1608 tal. 7 ma. 5 cand.

3. How many dollars will pay for an invoice of tea, amounting to 6446 tales 1 mace 6 candareens?

72) 6446 1 6 (8953
576

686
648

381
360

216
216   Ans. 8953 dols.

MANILLA.

Accounts are kept in Dollars, Reals, and Quartos.

12 quartos make 1 real ; 8 reals 1 dollar.

The arobe is 25 lb. 5½ arobes make 1 picul.
Their 100 lb. is equal to 104 lb. English.

1. What will 1897 bags of sugar amount to, weighing neat 1361 piculs 1 arobe 17½ lb. at 6 dollars per arobe ?

$$
\begin{array}{r}
1361\ \ 1\ \ 17\tfrac{1}{2} \\
6 \\
\hline
8166
\end{array}
$$

1 ar.	⅕	1 1½
12½ lb.	½	4¾
5	⅕	1¾

$$
\overline{8168\ \ 0} \qquad \text{Ans. 8168 dollars.}
$$

       pic. ar. lb. dol. re. dol. re.
2. 118 bags of sugar, weighing 89 1 22½ at 5 7   Ans. 524 6
3. 663 do.····do.········· 469 3 18 ·· 6 ···· 2819

COLUMBO, ISLE OF CEYLON.

The money is in paper, silver, and gold.

Paper money is in the bills of the Company, and is of uncertain value.

Silver is in the rupees of different parts of India.

The Sicca rupee is worth more than any other by 7 to 8 per cent.

Gold is the Mohur pagoda.

The exchange is various, as silver is rarely seen.

6 stivers	····make····	1 shilling Flemish.
8 shillings	············	1 rix dollar.
30 stivers	············	1 rupee.
64½ do.	············	1 Spanish dollar.

EXCHANGE. 217

JAPAN.

Accounts are in Tales, Mace, and Candareens.

10 candareens make 1 mace.
10 mace............1 tale = $\frac{3}{4}$ of a dollar, or 75 cents.

Ten mace are equal to 1 rix dollar.

Six tales make a corban, a gold coin not used in accounts.

In Weights—10 tales make 1 mace ; 16 mace 1 catta.

The ichan or hickey is $3\frac{1}{2}$ feet.
The balee is 65 quarts.

Thirty-five per cent. was the duty on privileged imports in 1799. It is on the exports (which are all free of duty) that the Dutch make their profit upon their return to Batavia. A privilege is granted to the captain of the Dutch ships to carry money, which often sells at an advance.

How much is the neat proceeds of 4 silver watches, at 35 tales each, deducting the duty of 35 per cent. ?

```
         35 tales.
          4
        ─────
         140
         35 per cent.
        ─────
         700          Sales  140
         420          Duty    49
        ─────
         49,00   Ans. neat proceeds 91 tales.
```

FORM OF AN ACCOUNT OF SALES.

	tales.	Duties. tales.	Neat. tales.
4 silver watches, 1st kind	35	49	91
6 silver watches, 2d kind	23,1	48,5,1	90,0,9

The article is given in the first column, the price in the next column, the duties in the third, and the neat proceeds in the fourth.

T

PARTICULARS

Of the TONNAGE *of* GOODS, *as calculated to make up the Tonnage for the Freight of Goods, brought in East-India or China ships to Europe*—viz.

PIECE GOODS.

FORT ST. GEORGE.	Pieces to the Ton.	BENGAL.	Pieces to the Ton.
ALLEIARS	800	Elatches	R.800
Betelles	400	Emmerties	600
Callawapores	800	Gurrahs	400
Chintz of all sorts	R.400	Ditto, long	200
Ginghams	800	Ginghams, coloured	600
Izzarees	800	Humhums	400
Longcloths	160	Habassies	600
Moorees	800	Humhums, quilted	100
Sallampores	400	Jamdannies	800
Succundies	800	Jamwars	600
		Laccowries	600
BENGAL.		Lungees Herba	800
Addatios	700	Mulmuls	400
Allibalhes	400	Ditto handkerchiefs	400
Allachaws	1200	Mahamodietes	400
Allibannies	R.800	Mamodies	R.400
Arras	R.400	Nillaes	800
Atchabannies	800	Nainsooks	400
Baftaes	R.400	Peniascoes	800
Bandannoes, or Taffa de Foolas	R.800	Photaes	R.800
Carridarries	600	Percaulas	800
Callipatties	400	Putcahs	R.400
Coopees	600	Romals	R.800
Callicoes	400	Sannoes	400
Chillaes	600	Seerhetties	400
Chowtars	600	Seerbands	600
Chunderbannies	800	Seersuckers	600
Chinnachures	R.800	Seerhaudconnaes	400
Cambrics	R.400	Seershauds	R.400
Chucklaes	400	Seerbafts	400
Cushtaes	800	Shaulbafts	400
Cossaes	400	Succatoons	R.800
Charconnaes	600	Sooseys	400
Cuttannaes	R 800	Sorts	400
Doosooties	R.400	Taffeties of all sorts	R.800
Dungaries	R.400	Tanjeebs	400
Doreas	400	Tepoys	R.800
Dimities	600	Terrindams	400
Diapers, broad	400	Tainsooks	400
Ditto, narrow	600		

EXCHANGE.

PIECE GOODS.

BOMBAY.

	Pieces to the Ton.
Annabatches	R.400
Bombay stuffs	R.400
Byrampauts	400
Bejutapauts	R.400
Boralchawders or brawls	1200
Betellees	400
Chelloes	R.400
Chintz of all sorts	R.400
Dooties	R.400
Guinea stuffs, large	600
Ditto, small	1200
Longcloths, whole pieces	160
Ditto, half ditto	320
Lemances	R.800
Musters	400
Nunsarees	R.400
Neganepauts	400
Niccanees, large	600
Ditto, small	600
Salampores	400
Stuffs, brown	R.400
Tapseils, large	400
Ditto, small	600

CHINA.

	Pieces to the Ton.
Nankeen cloth	R.400
Silks, of all sorts	R.800

China ware, 50 cubical feet to the ton, or about 4 chests of the usual dimensions.

Other measurable goods, 50 cubical feet to the ton.

N. B. Where the letter R. is set against pieces of 400 to the ton, it shews those goods are to be reduced, or brought to a standard of 16 yards long and 1 broad.

Where against pieces of 800 to the ton, to 10 yards long and 1 broad.

EXAMPLE.

1000 pieces of 12 yards long and 1⅛ broad, at 400 to the ton, make 844 pieces, or 2 tons 44 pieces.

1000 pieces of 10½ yards long and 1⅜ broad, at 800 to the ton, is 1181 pieces, or 1 ton 381 pieces.

WEIGHABLE GOODS.

	Cwt. to the Ton.		Cwt. to the Ton.
Arrangoes	20	Gum Opoponax	16
Aloes	16	....Sagapenum	18
Benjamin	20	....Sarcocol	18
Borax	20	Indigo	12
Cardemons, fine goods	12	Iron Kintlage	20
Cakelack	16	Musk	20
Carmenia Wool	10	Myrrh	16
Cambogium	20	Mother-of-Pearl Shells	20
Cassia Lignea	8	Nux Vomita	15
Cassia Buds	12	Pepper	16
Camphire	15	Quicksilver	20
Cotton Yarn, Fine Goods	10	Rhubarb	8
Cowries....Gruff ditto	20	Raw Silk	10
Coffee......Fine do.	18	Ditto in chests	8
Cinnabar	10	Ditto in bales or bundles	10
Cloves	12	Redwood	20
Dragon's Blood	20	Rice	20
Gum Arabic	16	Shellack	16
....Elemi	16	Seedlack	18
....Ammoniacum	16	Sticklack	16

EXCHANGE.

WEIGHABLE GOODS.

	Cwt. to the Ton.		Cwt. to the Ton.
Salt-Petre	20	Tea, Green	8
Senna	8	....Bohea	10
Sago	16	Arrack....Gauge gallons	251
Ditto, packed in China ware	—	Canes	Tale 300
Tutenague	20	Wanghees and Bamboes	3000
Turmeric	16	Rattans equal to 16 cwt.	6000
Tincal	16		

• • • • • • • • • • •

ARBITRATION OF EXCHANGE.

When the rates of exchange between several countries in succession are given, to find the rate of exchange between the first and last place in the correspondence.

Rule. Find by proportion the value of the sum originally remitted in the different monies of the countries through which it passes according to the rates of the different exchanges and so proceed till the whole is finished. Or,

Multiply all the first terms of the different statings together for a divisor, and the second terms, together with the sum remitted, for a dividend, and the quotient is the amount received in the denomination of the last place in the correspondence: from this result the rate of exchange is readily found by proportion.

Examples.

1. A merchant in London has credit for 500 piastres in Leghorn for which he can draw directly at 52d. sterling per piastre, but chusing to have it remitted by a circular rout, they are sent, by his order, to Venice at 95 piastres for 100 ducats banco; from thence to Cadiz at 350 maravedies per ducat banco; from thence to Lisbon at 630 reas per piastre of 272 maravedies; from thence to Amsterdam at 48d. Flemish for 400 reas; from thence to Paris at 54d. Flemish per crown; and from thence to London at 30d. sterling per crown: What is the arbitrated price between London and Leghorn per piastre, and what is gained or lost by this circular remittance without reckoning expences?

ARBITRATION OF EXCHANGE. 221

piast: d. ban.   piast.   d. ban.
 95 : 100   ::   500   : $526\frac{6}{19}$ in Venice.
 d. b. mar.   d. b.   mar.
  1 : 350   :: $526\frac{6}{19}$: $184210\frac{10}{19}$ in Cadiz.
 mar. reas.   mar.   reas.
 272 : 630   :: $184210\frac{10}{19}$: 426664 in Lisbon.
 reas. d. fl.   reas.   d. fl.
 400 :   48   ::   426664 : $51199\frac{1}{4}$ in Amsterdam.
 d. fl. cr.   d. fl.   cr.
  54 :   1   :: $51199\frac{1}{4}$: $948\frac{5}{36}$ in Paris.
 cr.   d. st.   cr.   £. s. d.
  1 : 30   :: $948\frac{5}{36}$: 118 10 $4\frac{1}{4}$ sterling.

Or thus,
 piast. d. b. mar. reas. d. fl. cr.
 $95 \times 1 \times 272 \times 400 \times 54 \times 1 = 55814400$
 piast. d. b. mar. reas. d. fl. cr. d. st.
 $500 \times 100 \times 350 \times 630 \times 48 \times 1 \times 30 = 1587600000000$

                             12
558144|00)1587600000|00) $28444\frac{1}{4}$
         1116288
         ———————   2|0)237|0 $4\frac{1}{4}$
          4713120
          4465152   £.118 10 $4\frac{1}{4}$ as above.
         ———————
          2479680
          2232576
         ———————
          2471040
          2232576
         ———————
          2384640   piast. £. s. d.   piast. d.
          2232576   500 : 118 10 $4\frac{1}{4}$ :: 1 : $56\frac{1777}{2000}$
         ———————
           152064
                4
         ———————
 558144)608256($\frac{1}{4}$
        558144
         ———————
          50112   £. s. d.
Amount received by circular remittance 118 10 $4\frac{1}{4}$
500 piastres at 52d. 108 6 8
                                        ——————————
Ans. { Gained by circular remittance ..... £.10 3 $8\frac{1}{4}$
     { Arbitrated value of a piastre by ditto $56\frac{1777}{2000}$d. st.
T 2

ARBITRATION OF EXCHANGE.

2. A merchant in Boston has £.225 sterling in London which he can draw for at 54*d.* sterling per dollar, but chusing to try a circular rout it is sent to Dublin at £.100 sterling for £.109 Irish; thence to Hamburgh at 12½ marks banco per pound Irish; thence to Amsterdam at 33 florins for 40 marks banco; thence to Copenhagen at 5 florins for 2 rix dollars of Denmark; thence to Bremen at 3 marks per rix dollar of Denmark; thence to Russia at 5 marks for 2 rubles; thence to Bordeaux at 5 francs per ruble; thence to Cadiz at 18 reals plate for 10 francs; thence to Lisbon at 1250 reals plate for 100 milreas; thence to Leghorn at 750 soldi for 88 millreas; thence to Smyrna at 2 soldi per piastre; thence to Jamaica at 24*d.* Jamaica currency per piastre; and thence to Boston at 80*d.* Jamaica currency per dollar: What is gained or lost by this circular remittance?

Ans. 117 dols. 42 cts. gained.

. . . .

AMERICAN DUTIES

ARE CALCULATED AS IN THE FOLLOWING.

EXAMPLES.

1. What is the duty on 2885 gallons of molasses, at 5 cts. per gallon?

```
    2885
       5
  ───────
  14425 cents.    Ans. 144 dols. 25 cts.
```

2. What is the duty on the above molasses, if imported in a foreign vessel, the rate being 5½ cents per gallon, or 10 per cent. more than an American vessel?

```
    2885                Or,     144,25 as above.
      5½              10 per cent.  14,42½
  ───────                       ─────────
   14425                 dols. 158,67½
    1442½
  ───────
  dols. 158,67½
```

Ans. 158 dols. 67½ cts.

AMERICAN DUTIES.

3. How much is the duty on 3720 gallons of gin, at $31\frac{9}{10}$ cents per gallon?

```
   3720            3720
   31 9/10            9
   ─────         ─────────
   3720         10)33480
  11160
   3348            3348
  ──────
dols. 1186,68      Ans. 1186 dols. 68 cents.
```

                                                                           dols. cts.

4. 1273 lb. chocolate at 3 cents ············ Ans. 38 19
5.   965 lb. do. in a foreign vessel at $3\frac{3}{10}$ do.···· 31 $84\frac{1}{2}$
6. 1149 lb. cheese at 7 ditto ················ 80 43
7. 1295 lb. do. in a foreign vessel at $7\frac{7}{10}$ do.···· 99 $71\frac{1}{2}$
8. 1879 gals. Champaign wine at 45 do.········ 845 55
9. 2675 do. London particular Madeira at 58 do. 1551 50

10. What is the duty on 53 cwt. 2 qrs. 21 lb. of untarred Cordage, at 225 cents per cwt.?

```
                   225
                    53
                  ─────
                   675
                  1125
  2 qrs.    ½     112½
  14 lb.    ¼      28
  7 do.     ⅛      14
                 ──────
                 120,79⅜     Ans. 120 dols. 79½ cts.
```

11. What is the duty on the above cordage in a foreign vessel, at $247\frac{1}{2}$ cts. per cwt.?

                                                   Ans. 132 dols. $87\frac{1}{4}$ cts.

AMERICAN DUTIES.

12. How much is the duty on 4 hhds. of brown sugar, wt. gross 38 cwt. 3 qrs. 19 lb. tare 12 lb. per 100, at 2½ cents per lb. ?

```
              3800
               456  = excess 12 per cent.
                84
                19
              ─────
       gross  4359.            4359
       tare    523.              12
              ─────           ──────
              3836            523,08
                2½
              ─────
              7672
              1918
              ─────
              95,90
```

        Ans. 95 dols. 90 cts.

13. What is the duty on this sugar, in a foreign vessel, at 2¾ cents per lb. ?

        Ans. 105 dols. 49 cts.

The mode of estimating ad valorem rates of duty.

The ad valorem rates of duty, upon goods, wares and merchandizes, at the place of importation, shall be estimated by adding 20 per cent. to the actual cost thereof, if imported from the Cape of Good Hope, or from any other place beyond the same, and 10 per cent. on the actual cost thereof, if imported from any other place or country, including all charges, commissions, outside packages and insurance excepted.—*(See Laws of the United States.)*

AMERICAN DUTIES.

EXAMPLES.

1. What is the duty on an invoice of silver and plated ware, imported from London, the cost exclusive of commissions, &c. being £.359 18 4, at 15 per cent. ad valorem?

```
                          359
                          444   cents per £. sterling.
                        ─────
                         1436
                         1436
                         1436
    10s.    ½            222
     5      ¼            111
     3  4d. ⅓             74
                        ─────
   actual cost       159803  cents
10 per cent. added    15980
                     ──────
                     175783
   10   1/10          17578
    5    ½             8789
                     ──────
  for 15 per cent.    26367  cents.
```

Ans. 263 dols. 67 cts.

2. What will it amount to in a foreign vessel, at 16½ per cent. ad valorem? Ans. 290 dols. 4 cents.

. . . .

The rates at which all foreign coins and currencies are estimated at the Custom-Houses of the United States.

	Dols.	Cts.
Each pound sterling of Great Britain, at	4	44
Each pound sterling of Ireland	4	10
Each livre tournois of France		18½
Each florin or gilder of the United Netherlands		40
Each mark banco of Hamburgh		33⅓
Each rix dollar of Denmark	1	
Each real of plate of Spain		10
Each real of vellon of Spain		5
Each milree of Portugal	1	24
Each tale of China	1	48
Each pagoda of India	1	84
Each rupee of Bengal		50

228   PROGRESSION.

2. A man being asked how many sons he had, said that the youngest was 4 years old, and the eldest 32, and that he increased one in his family every 4 years, how many had he?
Ans. 8.

The second, third and fourth given to find the first.

RULE. Multiply the fourth by the third, made less by 1 the product subtracted from the second gives the first.

EXAMPLES.

1. A man in 10 days went from Boston to a certain town in the country, every day's journey increasing the former by 4 and the last day he went was 46 miles, what was the first?
Ans. 10 miles.

$4 \times \overline{10-1} = 36$ then $46 - 36 = 10$, the first day's journey.

2. A man takes out of his pocket at 8 several times, so many different numbers of shillings, every one exceeding the former by 6; the last 46, what was the first?   Ans. 4.

The second, third and fifth given to find the first.

RULE. Divide the fifth by the third, and from the quotient subtract half the product of the fourth, multiplied by the third less 1, gives the first.

EXAMPLE.

A man is to receive £.360 at 12 several payments, each to exceed the former by £.4, and is willing to bestow the first payment on any one that can tell him what it is; what will that person have for his pains?   Ans. £.8.

$$360 \div 12 = 30 \text{ then } 30 - \frac{4 \times \overline{12-1}}{2} = 8, \text{ the first payment.}$$

The first, third and fourth given to find the second.

RULE. Subtract the fourth from the product of the third, multiplied by the fourth, that remainder added to the first gives the second.

Example.

What is the last number of an Arithmetical Progression, beginning at 6, and continuing by the increase of 8 to 20 places?
Ans. 158.

$20 \times 8 - 8 = 152$ then $152 + 6 = 158$, the last number.

* * * * * *

GEOMETRICAL PROGRESSION

Is the increasing or decreasing of any rank of numbers by some common ratio, that is, by the continual multiplication or division of some equal number: As 2, 4, 8, 16, increase by the multiplier 2, and 16, 8, 4, 2 decrease by the divisor 2.

NOTE. When any number of terms is continued in Geometrical Progression, the product of the two extremes will be equal to any two means, equally distant from the extremes: As 2, 4, 8, 16, 32, 64, where $64 \times 2 = 4 \times 32 = 8 \times 16 = 128$.

When the number of terms are odd, the middle term multiplied into itself will be equal to the two extremes, or any two means, equally distant from the mean: As 2, 4, 8, 16, 32, where $2 \times 32 = 4 \times 16 = 8 \times 8 = 64$.

In Geometrical Progression the same five things are to be observed as in Arithmetical, viz.
1. The first term.
2. The last term.
3. The number of terms.
4. The equal difference or ratio.
5. The sum of all the terms.

NOTE. As the last term in a long series of numbers, is very tedious to come at, by continual multiplication; therefore, for the reader finding it out, there is a series of numbers made use of in Arithmetical Proportion, called indices, beginning with an unit, whose common difference is one, whatever number of indices you make use of, set as many numbers (in such Geometrical Proportion as is given in the question) under them:

As 1, 2, 3, 4, 5, 6 indices.
   2, 4, 8, 16, 32, 64 numbers in Geometrical Proportion.

But if the first term in Geometrical Proportion be different from the ratio, the indices must begin with a cypher.

As 0, 1, 2, 3, 4, 5, 6 indices.
   1, 2, 4, 8, 16, 32, 64 numbers in Geometrical Proportion.

PROGRESSION.

When the indices begin with a cypher, the sum of the indices made choice of must be always one less than the number of terms given in the question, for 1 in the indices is over the second term, and 2 over the third, &c.

Add any two of the indices together, and that sum will agree with the product of their respective terms.

As in the first table of indices 2 + 5 = 7
Geometrical proportion .... 4 × 32 = 128

Then in the second
2 + 4 = 6
4 × 16 = 64

In any Geometrical Progression proceeding from unity, the ratio being known, to find any remote term, without producing all the intermediate terms.

RULE. Find what figures of the indices added together would give the exponent of the term wanted, then multiply the numbers standing under such exponent into each other, and it will give the term required.

NOTE. When the exponent 1 stands over the second term, the number of exponents must be 1 less than the number of terms.

EXAMPLES.

1. A man agrees for 12 peaches, to pay only the price of the last, reckoning a farthing for the first, a half-penny for the second, &c. doubling the price to the last, what must he give for them?

0, 1, 2, 3, 4, exponents.
1, 2, 4, 8, 16, number of terms.

16 = 4
16 = 4
―――
256 = 8
8 = 3

4 + 4 + 3 = 11, number of terms less 1.

4)2048 = 11 numb. farth.
―――
12)512
―――
20)42 8
―――
£.2 2 8 answer.

2. A country gentleman going to a fair to buy some oxen, meets with a person who had 23, he demanding the price of them, was answered £.16 apiece; the gentleman bids him £.15

PROGRESSION.

apiece, and he would buy all; the other tells him it would not be taken, but if he would give what the last ox would come to, at a farthing for the first, and doubling it to the last, he should have all. What was the price of the oxen?

Ans. £.4369 1s. 4d.

In any Geometrical Progression, not proceeding from unity, the ratio being given, to find any remote term, without producing all the intermediate terms.

RULE. Proceed as in the last, only observe that every product must be divided by the first term.

EXAMPLES.

1. A sum of money is to be divided among eight persons, the first to have £.20, the second £.60, and so on in triple proportion, what will the last have?

0. 1. 2. 3. $\dfrac{540 \times 540}{20} = 14580$ then $\dfrac{14580 \times 60}{20} = 43740$
20. 60. 180. 540.

Ans. £.43740.

$3 + 3 + 1 = 7$ one less than the number of terms.

2. A gentleman, dying, left 9 sons, to whom and to his executor, he bequeathed his estate in manner following: To his executor £.50; his youngest son was to have as much more as the executor, and each son to exceed the next younger by as much more; what was the eldest son's portion?

Ans. £.25600.

The first term, ratio, and number of terms given, to find the sum of all the terms.

RULE. Find the last term as before, then subtract the first from it, and divide the remainder by the ratio less one, to the product of which add the greater, and it gives the sum required.

EXAMPLES.

1. A servant skilled in numbers agreed with a gentleman to serve him 12 months, provided he would give him a farthing

for his first month's service, a penny for the second, and 4d. for the third, &c.—what did his wages amount to ?

$$256 \times 256 = 65536, \text{ then } 65536 \times 64 = 4194304$$

0. 1. 2. 3. 4.
1. 4. 16. 64. 256.
(4+4+3=11. No. of terms less 1.)

$$\frac{4194304-1}{4-1} = 1398101; \text{ then}$$

$$1398101 + 4194304 = 5592405 \text{ farthings.}$$

Ans. £.5825 8s. 5¼d.

2. A man bought a horse, and by agreement was to give a farthing for the first nail, three for the second, &c.; there were 4 shoes, and in each shoe 8 nails; what was the worth of the horse? Ans. £.965114681693 13s. 4d.

3. A certain person married his daughter on new-year's day, and gave her husband one shilling towards her portion, promising to double it on the first day of every month for one year; what was her portion? Ans. £.204 15s.

4. A laceman well versed in numbers, agreed with a gentleman to sell him 22 yards of rich gold brocaded lace, for 2 pins the first yard, 6 pins the second, &c., in triple proportion. I desire to know what he sold the lace for, if the pins were valued at 100 for a farthing; also, what the laceman got or lost by the sale thereof, supposing the lace stood him in £.7 per yard. Ans. The lace sold for £.326886 0s. 9d.
Gain £.326732 0s. 9d.

PERMUTATION

Is the changing or varying of the order of things.

RULE. Multiply all the given terms one into another, and the last product will be the number of changes required.

EXAMPLES.

1. How many changes may be rung upon 12 bells, and how long would they be ringing but once over, supposing 10 changes might be rung in one minute, and the year to contain 365 days 6 hours?

$1 \times 2 \times 3 \times 4 \times 5 \times 6 \times 7 \times 8 \times 9 \times 10 \times 11 \times 12 = 479001600$ changes, which $\div 10 = 47900160$ minutes, and if reduced is $= 91$ years 3 weeks 5 days and 6 hours.

2. A young scholar coming into a town for the conveniency of a good library, demands of a gentleman with whom he lodged, what his diet would cost for a year, who told him £.10; but the scholar, not being certain what time he should stay, asked him what he must give him for so long as he could place his family (consisting of 6 persons besides himself) in different positions, every day at dinner; the gentleman, thinking it could not be long, tells him £.5, to which the scholar agrees: what time did the scholar stay with the gentleman?

Ans. 5040 days.

EXTRACTION OF THE SQUARE ROOT.

EXTRACTING THE SQUARE ROOT is to find out such a number as being multiplied into itself, the product will be equal to the given number.

RULE. 1. Point the given number, beginning at the unit's place, then to the hundred's, and so upon every second figure throughout.

2. Seek the greatest square number in the first point, towards the left hand, placing the square number under the first point, and the root thereof in the quotient; subtract the square number from the first point, and to the remainder bring down the next point and call that the resolvend.

3. Double the quotient, and place it for a divisor on the left hand of the resolvend; seek how often the divisor is contained in the resolvend (reserving always the unit's place) and put the answer in the quotient, and also on the right hand side of the divisor; then multiply by the figure last put in the quotient, and subtract the product from the resolvend; bring down the next point to the remainder (if there be any more) and proceed as before.

Roots.   1. 2. 3. 4. 5. 6. 7. 8. 9.
Squares. 1. 4. 9. 16. 25. 36. 49. 64. 81.

EXTRACTION OF THE SQUARE ROOT.

EXAMPLES.

1. What is the square root of 119025?

```
119025(345
  9
  ──
64)290
   256
   ───
  685)3425
      3425
```
                                     Ans. 345.

2. What is the square root of 106929?   Ans. 327
3. What is the square root of 2268741? Ans. 1506,23+
4. What is the square root of 7596796? Ans. 2756,228+
5. What is the square root of 36372961? Ans. 6031
6. What is the square root of 22071204? Ans. 4698

When the given number consists of a whole number and decimals together, make the number of decimals even, by adding cyphers to them, so that there may be a point fall on the unit's place of the whole number.

7. What is the square root of 3271,4007? Ans. 57,19+
8. What is the square root of 4795,25731? Ans. 69,247+
9. What is the square root of 4,372594?   Ans. 2,091+
10. What is the square root of 2,2710957? Ans. 1,50701+
11. What is the square root of ,00032754? Ans. ,01809+
12. What is the square root of 1,270054? Ans. 1,1269+

To extract the square root of a vulgar fraction.

RULE. Reduce the fraction to its lowest terms, then extract the square root of the numerator for a new numerator, and the square root of the denominator for a new denominator.

If the fraction be a surd, (i. e.) a number whose root can never be exactly found, reduce it to a decimal, and extract the root from it.

EXAMPLES.

13. What is the square root of $\frac{2304}{3184}$?   Ans. $\frac{4}{5}$.
14. What is the square root of $\frac{2704}{4225}$?   Ans. $\frac{4}{5}$.
15. What is the square root of $\frac{2834}{4225}$?   Ans. $\frac{8}{9}$.

EXTRACTION OF THE SQUARE ROOT.

SURDS.

16. What is the square root of $\frac{275}{341}$?   Ans. ,89802+
17. What is the square root of $\frac{367}{475}$?   Ans. ,86602+
18. What is the square root of $\frac{478}{549}$?   Ans. ,93308+

To extract the square root of a mixed number.

RULE. 1. Reduce the fractional part of the mixed number to its lowest term, and then the mixed number to an improper fraction.

2. Extract the roots of the numerator and denominator for a new numerator and denominator.

If the mixed number given be a surd, reduce the fractional part to a decimal, annex it to the whole number, and extract the square root therefrom.

EXAMPLES.

19. What is the square root of $51\frac{21}{25}$?   Ans. $7\frac{1}{5}$.
20. What is the square root of $27\frac{9}{16}$?   Ans. $5\frac{1}{4}$.
21. What is the square root of $9\frac{4}{9}$?   Ans. $3\frac{1}{7}$.

SURDS.

22. What is the square root of $85\frac{1}{5}$?   Ans. 9,27+
23. What is the square root of $8\frac{4}{7}$?   Ans. 2,9519+
24. What is the square root of $6\frac{2}{5}$?   Ans. 2,5298+

THE APPLICATION.

1. There is an army consisting of a certain number of men, who are placed rank and file, that is, in the form of a square, each side having 576 men, I desire to know how many the whole square contains ?   Ans. 331776.

2. A certain pavement is made exactly square, each side of which contains 97 feet, I demand how many square feet are contained therein ?   Ans. 9409.

To find a mean proportional between any two given numbers.

RULE. The square root of the product of the given number is the mean proportional sought.

EXTRACTION OF THE SQUARE ROOT.

EXAMPLES.

1. What is the mean proportional between 3 and 12?
Ans. $3 \times 12 = 36$ then $\sqrt{36} = 6$ the mean proportional.

2. What is the mean proportional between 4276 and 842?
Ans. 1897,4 +.

To find the side of a square equal in area to any given superfices.

RULE. The square root of the content of any given superfices, is the square equal sought.

EXAMPLES.

3. If the content of a given circle be 160, what is the side of the square equal? Ans. 12,64911.

4. If the area of a circle is 750, what is the side of the square equal? Ans. 27,38612.

The area of a circle given to find the diameter.

RULE. As 355 : 452, or as 1 : 1,273239 :: so is the area : to the square of the diameter;—or, multiply the square root of the area by 1,12837, and the product will be the diameter.

EXAMPLE.

5. What length of cord will fit to tie to a cow's tail, the other end fixed in the ground, to let her have liberty of eating an acre of grass, and no more, supposing the cow and tail to be 5 yards and a half? Ans. 6,136 perches.

The area of a circle given to find the periphery, or circumference.

RULE. As 113 : 1420, or as 1 : 12,56637 :: the area : to the square of the periphery, or multiply the square root of the area by 3,5449, and the product is the circumference.

EXAMPLES.

6. When the area is 12, what is the circumference?
Ans. 12,2798.

7. When the area is 160, what is the periphery?
Ans. 44,84.

Any two sides of a right angled triangle given to find the third side.

1. The base and perpendicular given to find the hypothenuse.

RULE. The square root of the sum of the squares of the base and perpendicular is the length of the hypothenuse.

EXAMPLES.

8. The top of a castle from the ground is 45 yards high, and is surrounded with a ditch 60 yards broad; what length must a ladder be to reach from the outside of the ditch, to the top of the castle? Ans. 75 yards.

```
                              │45 yards.
                              │Height of the Castle.
                              │Perpendicular.
      Ditch.                  │
    Base 60 yards.
```

9. The wall of a town is 25 feet high, which is surrounded by a moat of 30 feet in breadth, I desire to know the length of a ladder that will reach from the outside of the moat to the top of the wall. Ans. 39,05 feet.

· · · ·

The hypothenuse and perpendicular given to find the base.

RULE. The square root of the difference of the squares of the hypothenuse and perpendicular is the length of the base.

The base and hypothenuse given to find the perpendicular.

RULE. The square root of the difference of the hypothenuse and base is the height of the perpendicular.

N. B. The two last questions may be varied for examples to the two last propositions.

* * * *

Any number of men being given to form them into a square battle, or to find the number of ranks and files.

RULE. The square root of the number of men given, is the number of men either in rank or file.

10. An army consisting of 331776 men, I desire to know how many in rank and file?   Ans. 576.

11. A certain square pavement contains 48841 square stones, all of the same size, I demand how many are contained in one of the sides.   Ans. 221.

* * * * * *

EXTRACTION OF THE CUBE ROOT.

To extract the Cube Root is to find out a number which being multiplied into itself, and then into that product, produceth the given number.

RULE 1. Point every third figure of the cube given, beginning at the unit's place, seek the greatest cube to the first point and subtract it therefrom, put the root in the quotient, and bring down the figures in the next point to the remainder for a resolvend.

2. Find a divisor by multiplying the square of the quotient by 3. See how often it is contained in the resolvend, rejecting the units and tens, and put the answer in the quotient.

3. To find the *subtrahend*. 1. Cube the last figure in the quotient. 2. Multiply all the figures in the quotient by 3, except the last, and that product by the square of the last. 3. Multiply the divisor by the last figure. Add these products together, gives the subtrahend, which subtract from the resolvend; to the remainder bring down the next point and proceed as before.

ROOTS. 1. 2. 3. 4. 5. 6. 7. 8. 9.
CUBES. 1. 8. 27. 64. 125. 216. 343. 512. 729.

CUBE ROOT.

Example.

What is the cube root of 99252847?

```
99252847(463
64 =Cube of 4.
```
Divisor.─────
Square of 4×3=48)35252 Resolvend

```
       216 =Cube of 6
       432 =4×3× by square of 6
       288 =Divisor × by 6
     ─────
     33336 Subtrahend
```
Divisor.─────
Sq. of 46×3=6348)1916847 Resolvend

```
        27 =Cube of 3
      1242 =46×3× by square of 3
     19044 =Divisor × by 3
     ─────
   1916847 Subtrahend.
```

. . . .

Another new and more concise method of extracting the Cube Root.

RULE. 1. Point every third figure of the cube given, beginning at the unit's place, then find the nearest cube to the first point, and subtract it therefrom, put the root in the quotient, bring down the figures in the next point to the remainder for a resolvend.

2. Square the quotient and triple the square for a divisor—as, 4×4×3=48. Find how often it is contained in the resolvend, rejecting units and tens, and put the answer in the quotient.

3. Square the last figure in the quotient, and put it on the right hand of the divisor:

As 6×6=36 put to the divisor 48=4836.

4. Triple the last figure in the quotient, and multiply by the former, put it under the other, units under the tens, add them together, and multiply the sum by the last figure in the quotient, subtract that product from the resolvend, bring down the next point and proceed as before.

CUBE ROOT.

EXAMPLES.

1. What is the cube root of 99252847?

```
Square of 4 × 3 = 48 divisor        99252847 (463
Square of 6 put to 48 = 4836        64
        6 × 3 × 4 =   72            ―――――
                    ―――             35252
             5556 ×  6 =            33336
Square of 46 = 2116 × 3 = 6348 divisor ―――――
Square of 3 = 9 put to 6348 = *634809   1916847
        3 × 3 × 46 =    414
                      ―――――
              638949 × 3 = 1916847
```

2. What is the cube root of 389017? Ans. 73.
3. What is the cube root of 5735339? Ans. 179.
4. What is the cube root of 32461759? Ans. 319.
5. What is the cube root of 84604519? Ans. 439.
6. What is the cube root of 259694072? Ans. 638.
7. What is the cube root of 48228544? Ans. 364.
8. What is the cube root of 27054036008? Ans. 3002.
9. What is the cube root of 22069810125? Ans. 2805.
10. What is the cube root of 122615327232? Ans. 4968.
11. What is the cube root of 219365327791? Ans. 6031.
12. What is the cube root of 673373097125? Ans. 8765.

When the given number consists of a whole number and decimal together, make the number of decimals to consist of 3, 6, 9, &c. places, by adding cyphers thereto, so that there may be a point fall on the unit's place of the whole number.

13. What is the cube root of 12,977875? Ans. 2,35
14. What is the cube root of 36155,027576? Ans. 33,06+
15. What is the cube root of ,001906624? Ans. ,124
16. What is the cube root of 33,230979637? Ans. 3,215+
17. What is the cube root of 15926,972504? Ans. 25,16+
18. What is the cube root of ,053258279? Ans. ,376+

••

* When the quotient is 1, 2, or 3, there must be a cypher put to supply the place of tens.

CUBE ROOT.

To extract the cube root of a vulgar fraction.

RULE. Reduce the fraction to its lowest terms, then extract the cube root of the numerator and denominator for a new numerator and denominator; but if the fraction be a surd, reduce it to a decimal, and then extract the root from it.

EXAMPLES.

19. What is the cube root of $\frac{250}{686}$?   Ans. $\frac{5}{7}$.
20. What is the cube root of $\frac{324}{1500}$?   Ans. $\frac{3}{5}$.
21. What is the cube root of $\frac{1520}{5130}$?   Ans. $\frac{2}{3}$.

SURDS.

22. What is the cube root of $\frac{4}{7}$?   Ans. ,829+
23. What is the cube root of $\frac{5}{9}$?   Ans. ,822+
24. What is the cube root of $\frac{2}{3}$?   Ans. ,873+

To extract the cube root of a mixed number.

RULE. Reduce the fractional part to its lowest terms, and then the mixed number to an improper fraction, extract the cube roots of the numerator and denominator for a new numerator and denominator; but if the mixed number given be a surd, reduce the fractional part to a decimal, annex it to the whole number, and extract the root therefrom.

EXAMPLES.

25. What is the cube root of $12\frac{1}{2}\frac{7}{7}$?   Ans. $2\frac{1}{3}$.
26. What is the cube root of $31\frac{15}{343}$?   Ans. $3\frac{1}{7}$.
27. What is the cube root of $405\frac{28}{125}$?   Ans. $7\frac{2}{5}$.

SURDS.

28. What is the cube root of $7\frac{1}{5}$?   Ans. 1,93+
29. What is the cube root of $9\frac{1}{6}$?   Ans. 2,092+
30. What is the cube root of $8\frac{4}{7}$?   Ans. 2,057+

THE APPLICATION.

1. If a cubical piece of timber be 47 inches long, 47 inches broad, and 47 inches deep, how many cubical inches doth it contain ?   Ans. 103823.

2. There is a cellar dug that is 12 feet every way, in length, breadth, and depth, how many solid feet of earth were taken out of it ?   Ans. 1728.

X

3. There is a stone of a cubic form, which contains 389017 solid feet, what is the superficial content of one of its sides?
Ans. 5329.

Between two numbers given, to find two mean proportionals.

RULE. Divide the greater extreme by the lesser, and the cube root of the quotient multiplied by the lesser extreme gives the lesser mean; multiply the said cube root by the lesser mean, and the product will be the greater mean proportional.

EXAMPLES.

4. What are the two mean proportionals between 6 and 162?
Ans. 18 and 54.
5. What are the two mean proportionals between 4 and 108?
Ans. 12 and 36.

To find the side of a cube that shall be equal in solidity to any given solid, as a globe, cylinder, prism, cone, &c.

RULE. The cube root of the solid content of any solid body given is the side of the cube of equal solidity.

EXAMPLE.

6. If the solid content of a globe is 10648, what is the side of a cube of equal solidity?
Ans. 22.

The side of the cube being given, to find the side of that cube, that shall be double, treble, &c. in quantity to the given cube.

RULE. Cube the side given, and multiply it by 2, 3, &c. the cube root of the product is the side sought.

EXAMPLE.

7. There is a cubical vessel, whose side is 12 inches, and it is required to find the side of another vessel that is to contain three times as much?
Ans. 17,306.

BIQUADRATE ROOT.

EXTRACTION OF THE BIQUADRATE ROOT.

To extract the Biquadrate Root is to find out a number, which being involved four times into itself, will produce the given number.

RULE. First extract the square root of the given number, then extract the square root of that square root, and it will give the biquadrate root required.

EXAMPLES.

1. What is the biquadrate of 27 ?   Ans. 531441.
2. What is the biquadrate of 76 ?   33362176.
3. What is the biquadrate of 275 ?   5719140625.

4. What is the biquadrate root of 531441 ?   27.
5. What is the biquadrate root of 33362176 ?   76.
6. What is the biquadrate root of 5719140625 ?   275.

A GENERAL RULE

FOR EXTRACTING THE ROOTS OF ALL POWERS.

1. PREPARE the number given for extraction, by pointing off from the unit's place, as the root required directs.
2. Find the first figure in the root, by the table of powers, which subtract from the given number.
3. Bring down the first figure in the next point to the remainder, and call it the dividend.
4. Involve the root into the next inferior power to that which is given; multiply it by the given power, and call it the divisor.
5. Find a quotient figure by common division, and annex it to the root; then involve the whole root into the given power, and call that the subtrahend.
6. Subtract that number from as many points of the given power as is brought down, beginning at the lowest place, and to the remainder bring down the first figure of the next point for a new dividend.
7. Find a new divisor, and proceed in all respects as before.

RULE FOR EXTRACTING, &c.

EXAMPLES.

1. What is the square root of 141376?

```
141376(376
  9
  ─────
6)51 dividend            3×2=6   divisor
                         37×37=1369 subtrahend
  1369 subtrahend        37×2=74  divisor
  ─────                  376×376=141376 subtrahend
74)  447 dividend

  141376 subtrahend                    Ans. 376.
```

2. What is the cube root of 53157376?

```
53157376(376
 27
 ─────
27)261 dividend          3×3×3=27 divisor
                         37×37×37=50653 subtrahend
  50653 subtrahend       37×37×3=4107 divisor
  ─────                  376×376×376=53157376 subtrahend
4107)25043 dividend

  53157376 subtrahend                  Ans. 376.
```

3. What is the biquadrate root of 19987173376?

```
        19987173376(376
         81
         ─────
       108)1188 dividend

          1874161 subtrahend

       202612) 1245563 dividend

          19987173376 subtrahend
```

3 × 3 × 3 × 4 =108 divisor
37× 37× 37× 37=1874161 subtrahend
37× 37× 37× 4 =202612 divisor
376×376×376×376=19987173376 subtrahend

Ans. 376.

DUODECIMALS.

DUODECIMALS, or Cross Multiplication, is a rule made use of in measuring and computing the dimensions of the several parts of buildings; it is likewise used to find ships' tonnage and the contents of bales, cases, &c.

Dimensions are taken in feet, inches, and parts.

Artificers' work is computed by different measures, viz.
Glazing, and masons' flat work, by the foot;
Painting, paving, plastering, &c. by the yard.
Partitioning, flooring, roofing, tiling, &c. by the square of 100 ft.
Brick-work, &c. by the rod of $16\frac{1}{2}$ feet, whose square is $272\frac{1}{4}$.

The contents of bales, cases, &c. by the ton of 40 cubic feet.
The tonnage of ships, by the ton of 95 feet.

RULE FOR MULTIPLYING DUODECIMALLY.

1. Under the multiplicand write the corresponding denominations of the multiplier.

2. Multiply each term in the multiplicand, (beginning at the lowest) by the feet in the multiplier; write each result under each respective term, observing to carry an unit from each lower denomination to its superior.

3. In the same manner, multiply the multiplicand by the inches in the multiplier, and write the result of each term, one place more to the right hand of them, in the multiplicand.

4. Work in the same manner with the other parts in the multiplier, setting the result of each term two places to the right hand of those in the multiplicand, and so on for thirds, fourths, &c.

5. Proceed in the like manner with all the rest of the denominations, and their sum will give the answer required.

DUODECIMALS.

EXAMPLES.

1. Multiply 4 feet 9 inches by 8 inches.

```
    4.  9
        8
    -------
    3.  2.
```
      Ans. 3 feet 2 inches.

2. Multiply 9 feet 6 inches by 4 feet 9 inches.

```
                    ft.   in.
                     9    6
                     4    9
                    -----------
ft. in.
 9  6 × 4 feet = 38   0
 9  6 × 9 inc. =  7   1   6
                 ---------------
                  45   1.  6
```
   Ans. 45 feet 1 inch and 6 twelfths.

3. What is the price of a marble slab, whose length is 5 feet 7 inches, and breadth 1 foot 10 inches, at 1 dollar per foot?
      Ans. 10 dols. 23 cents.

4. There is a house with three tiers of windows, 3 in a tier, the height of the first tier is 7 feet 10 inches, of the second 6 feet 8 inches, and of the third 5 feet 4 inches, and the breadth of each is 3 feet 11 inches; what will the glazing come to, at 14d. per foot? Ans. £.13 11s. 10½d.

5. If a house measures within the walls 52 feet 8 inches in length, and 30 feet 6 inches in breadth, and the roof be of a true pitch or the rafters ¾ of the breadth of the building, what will it come to roofing at 10s. 6d. per square?
      Ans. £.12 12s. 11¾d.

DUODECIMALS.

APPLICATION OF DUODECIMALS.

To find how many cubic or solid square feet (in order to ascertain the freight) are contained in cases, bales, &c. that is, how many cubic feet they will take up in a ship.

EXAMPLES.

1. Suppose the dimensions of a bale to be 7 feet 6 inches, feet 3 inches, and 1 foot 10 inches ; what is the solid content.?

```
                    ft.   in.
                     7     6
                     3     3
        ft.  in.    ─────────
         7   6 ×3 ft. = 22   6
         7.  6 ×3 in. =  1  10   6
                       ─────────────
                        24   4   6
                         1  10
ft.  in.  tw.          ─────────────
 24   4   6 ×1 ft. = 24    4   6
 24   4   6 ×10 in.= 20    3   9
                     ─────────────
                      44   8   3
```
Ans. 44 feet 8 inches and 3 twelfth parts.

2. What is the freight of a bale containing 65 feet 9 inches, at 15 dollars per ton of 40 feet ?

```
            dols. cts.                      decimally.
          15,00 for 40 feet                    65,75
                                                  15
 20 ft.  ½    7,50                           ───────
  5 ft.  ¼    1,87,5                           32875
  6 in.  1/10   ,18,7                           6575
  3      ½     ,09,3                         ───────
              ──────                        40)986,25
              24,65,5                         ──────
                                              24,65,6
```
Ans. 24 dols. 65½ cts.

3. A merchant imports from London 6 bales of the following dimensions, viz.

	Length. ft. in.	Height. ft. in.	Depth. ft. in.
No. 1.	2 10	2 4	1 9
2.	2 10	2 6	1 3
3.	3 6	2 2	1 8
4.	2 10	2 8	1 9
5.	2 10	2 6	1 9
6.	2 11	2 8	1 8

What are the solid contents, and how much will the freight amount to, at 20 dollars per ton?

The contents are, viz.	ft.	in.	Feet.
No. 1.	11	7	71,58
2.	8	10	20 dols. per ton.
3.	12	7	
4.	13	2	
5.	12	5.	40)1431,60
6.	13	0	
	71	7	35,79

Ans. 35 dols. 79 cts.

* * * * *

To find Ships' Tonnage by Carpenters' Measure.

RULE. For single decked vessels, multiply the length, breadth at the main beam, and depth of the hold together, and divide the product by 95.

EXAMPLE.

What is the tonnage of a single decked vessel, whose length is 60 feet, breadth 20 feet, and depth 8 feet.

```
    60 length
    20 breadth
   ----
  1200
     8 depth
   ----
95)9600(101 5/95
   95
   ---
   100
    95
   ---
     5.
```

Ans. $101\frac{5}{95}$ tons.

This is the usual method of tonnaging a single-decked vessel, having the deck bolted to the wale. But if it be required that the deck be bolted at any height above the wale, the custom is to pay the carpenter for one half of the additional height, to which the deck may be thus raised; that is, one half of the difference being added to the former depth gives the depth to be used in calculating the tonnage.

DUODECIMALS.

EXAMPLE.

A merchant, after having contracted with a carpenter to build a single-decked vessel of 60 feet keel, 20 feet beam, and 8 feet hold, desires that the deck be laid for 10 feet hold; required the tonnage to be paid for?

```
                    60 length
                    20 breadth
                   ─────
                  1200
1 = ⅕ diff. of depth + 8 = 9

            95)10800(113 55/95
               95
              ────
               130
                95
              ────
               350
               285
              ────
                65        Ans. 113 65/95 tons.
```

......

RULE. For a double-decked vessel, take half the breadth of the main beam for the depth of the hold, and work as for a single-decked vessel.

EXAMPLES.

1. What is the tonnage of a double-decked vessel, whose length is 65 feet, and breadth 21 feet 6 inches?

```
                       65   length
                       21 6 breadth
                      ────
                       65
                      130
  65 ft. × 6 in. =     32 6
                      ─────
                     1397 6
  ft.  in.            10 9  depth
 1397 6×10 ft. = 13975 0
 1397 6× 9 in. =  1048 1
                 ─────────
             95)15023 1(158 13/95
                 95
                ────
                552
                475
                ────
                773
                760
                ────
                 13          Ans. 158 13/95 tons.
```

250 DUODECIMALS.

The preceding question may be wrought thus:

```
              65
              21 6
            ———————
              65
             130
      6  ½  ————————
             1365
               32 6
            ————————
             1397 6
               10 9
            ————————
            13975 0
      6  ½    698 9
      3  ¼    349 4
            ————————
      95)15023 1  as before.
            ————————
             158 13/95 tons.
```

2. What will the above tonnage amount to, at 16 dols. per ton?

```
                              dols.
     158                       16
      16                       13
    ——————                   ——————
     948                       48
     158                       16
       2,18                 ——————
    ——————                  95)208(2,18
    2530,18                    190
                             ——————
                               180
                                95
                             ——————
                               850
                               760
                             ——————
Ans. 2530 dols. 18 cents.       90
```

3. Required the tonnage of a ship of 74 feet keel, and 26 ft 6 inches beam? Ans. 273 48/95 tons.

DUODECIMALS.

To find the Government Tonnage.

"If the vessel be double-decked, take the length thereof from the fore part of the main stem, to the after part of the stern post, above the upper deck; the breadth thereof at the broadest part above the main wales, half of which breadth shall be accounted the depth of such vessel, and then deduct from the length, three-fifths of the breadth, multiply the remainder by the breadth, and the product by the depth, and divide this last product by 95, the quotient whereof shall be deemed the true contents or tonnage of such ship or vessel; and if such ship or vessel be single-decked, take the length and breadth, as above directed, deduct from the said length three-fifths of the breadth, and take the depth from the under side of the deck plank, to the ceiling in the hold, then multiply and divide as aforesaid, and the quotient shall be deemed the tonnage."

EXAMPLES.

1. What is the government tonnage of a single-decked vessel, whose length is 69 feet 6 inches, breadth 22 feet 6 inches, and depth 8 feet 6 inches?

```
                    ft. in.
                    69  6 length.      22  6 breadth
        deduct      13  6 for ⅗ breadth.   3
                    ─────                ─────
                    56  0              5)67  6
                    22  6 breadth        ─────
                    ─────               13  6
                   112  0
                   112
   6 in.   ½        28  0
                   ──────
                  1260  0
                     8  6 depth
                   ──────
                 10080  0
   6 in.   ½       630  0
                   ──────
           95)10710 0(112 70/95 tons.
              95
              ───
              121
               95
              ───
              260
              190
              ───
               70                  Ans. 112 70/95 tons.
```

DUODECIMALS.

2. What is the government tonnage of a double-decked vessel, of the following dimensions, length 75 feet 6 inches, breadth 23 feet 4 inches, and depth 11 feet 8 inches?

```
            ft. in.
            75  6
            14  0 for ⅕ breadth         Or,         ft. in.
                                                    75  6
            ─────                                   14  0
            61  6                                   ─────
            23  4 breadth                           61  6
                                                    23  4
            ─────
            183                        61 ft. × 23 ft. = 1403  0
            122                         6 in. × 23 ft. =   11  6
   6 in. ½   11  6                 61 ft. 6 in. ×  4 in. =   20  6
   4 in. ⅓   20  6
                                                    1435  0
            ─────                                     11  8
            1435  0
              11  8 depth                           15785
                                   1435 ft. × 8 in. =  956  6
            ─────
            15785  0                                16741  8 as before.
   6 in. ½    717  6
   2 in. ⅓    239  2
            ─────
         95)16741  8(176 ⁶³⁄₉₅ tons.
            95
            ─────
            724
            665
            ─────
            591
            570
            ─────
             21                    Ans. 176⁶³⁄₉₅ tons.
```

3. What is the government tonnage of a double-decked vessel, of the following dimensions, length 82 feet 3 inches, breadth 24 feet 3 inches, and depth 12 feet 1½ inches?

Ans. 209⁶³⁄₉₅ tons.

TABLES OF *CORDAGE.*

A Cordage Table, *shewing how many fathoms, feet, and inches of a rope, of any size, not more than 14 inches, make a hundred weight; with the use of the table.*

Inches.	Fathoms.	Feet.	Inches.	Inches.	Fathoms.	Feet.	Inches.	Inches.	Fathoms.	Feet.	Inches.	Inches.	Fathoms.	Feet.	Inches.
1	486	0	0	4¼	26	5	3	7½	8	4	0	10¾	4	1	8
1¼	313	3	0	4½	24	0	0	7¾	8	3	6	11	4	0	3
1½	216	3	0	4¾	21	3	0	8	7	3	6	11¼	3	5	7
1¾	159	3	0	5	19	3	0	8¼	7	0	8	11½	3	4	1
2	124	3	0	5¼	17	4	0	8½	6	4	3	11¾	3	3	3
2¼	96	2	0	5½	16	1	0	8¾	6	2	1	12	3	2	3
2½	77	3	0	5¾	14	4	6	9	6	0	0	12¼	3	2	1
2¾	65	4	0	6	13	3	0	9¼	5	4	0	12½	3	2	0
3	54	0	0	6¼	12	2	0	9½	5	2	0	12¾	2	7	8
3¼	45	5	2	6½	11	3	0	9¾	5	0	6	13	2	5	3
3½	39	3	0	6¾	10	4	0	10	4	5	0	13¼	2	4	9
3¾	34	3	9	7	9	5	6	10¼	4	4	1	13½	2	4	0
4	30	1	6	7¼	9	1	6	10½	4	2	2	13¾	2	3	6
												14	2	2	1

USE OF THE TABLE.

At the top of the table, marked inches, fathoms, feet, inches, the first column is the thickness of the rope in inches and quarters, and the other three the fathoms, feet, and inches that make up a hundred weight of such a rope. One example will make it plain:

Suppose you desire to know how much of a seven-inch rope will make a hundred weight: Find 7 in the third column under inches, or thickness of rope, and against it in the fourth column you find 9 5 6, which shews that there will be 9 fathoms 5 feet 6 inches required to make one hundred weight.

A Table, *shewing the weight of any Cable or Rope of* 120 *fathoms in length, and for every half inch, from 3 to 24 inches in circumference.*

Inches.	Cwt. Qrs.	Inches.	Cwt. Qrs.	Inches.	Cwt. Qrs.	Inches.	Cwt. Qrs.	Inches.	Cwt. Qrs.
3	2 1	7	12 1	11	30 1	15½	60 0	20	100 0
3½	3 0	7½	14 0	11½	33 0	16	64 0	20½	105 0
4	4 0	8	16 0	12	36 0	16½	68 0	21	110 1
4½	5 0	8½	18 0	12½	39 0	17	72 1	21½	115 2
5	6 1	9	20 1	13	42 1	17½	76 2	22	121 0
5½	7 2	9½	22 2	13½	45 2	18	81 0	22½	126 2
6	9 0	10	25 0	14	49 0	18½	85 2	23	132 1
6½	10 2	10½	27 2	14½	52 2	19	90 1	23½	138 0
				15	56 1	19½	95 0	24	144 0

USE OF THE TABLE.

The first column marked for inches, is the thickness or circumference of the cable to every half inch from 3 to 24 inches; the second, marked Cwt. qrs. for the hundred weights and quarters that it will weigh if 120 fathoms in length.

For instance: Suppose it be a cable of 14½ inches; look against 14½ and you will find in the other column 52 cwt. 2 qrs. which shews that 120 fathoms of 14½ inch cable will weigh 52 cwt. 2 qrs. and so in others: and any quantity of a less length will weigh in proportion.

....

A ship was brought to anchor in a gale of wind, but the gale increasing, it was thought safest to cut the cables, in consequence of which 75 fathoms of 16 inches and 50 fathoms of 12 inches were lost; what must they be valued at in calculating the average; new cordage being then 14 dollars per cwt.?

CALCULATION.

120 fath. 16in. cable=64 cwt.	120 fath. 12 in. cab =36 cwt.
60 ....do......... 32	40 ......do........ 12
15 ....do......... 8	10 ......do........ 3
75 fath. weighing .. 40	50 fath. weighing .. 15
50 ......do 15	

dols. cts.
55 cwt. at 14 dols. per cwt..... 770 00
One third deducted for new.... 256 66⅔

Answer—dols. 513 33⅓

TABLES OF GOLD COIN.

A TABLE

For receiving and paying the Gold Coins of France *and* Spain, *at 100 cents for 27⅔ grains according to Act of Congress.*

grains.	dol.	cts.	137ths of a ct.	dwt.	dol.	cts.	137ths of a ct.	ounces.	dol.	cts.	137ths of a ct.
1	0	3	89	12	10	51	13	27	472	99	37
2	0	7	41	13	11	38	94	28	490	51	13
3	0	10	130	14	12	26	38	29	508	2	126
4	0	14	82	15	13	13	119	30	525	54	102
5	0	18	34	16	14	1	63	31	543	6	78
6	0	21	123	17	14	89	7	32	560	58	54
7	0	25	75	18	15	76	88	33	578	10	30
8	0	29	27	19	16	60	32	34	595	62	6
9	0	32	116	20	17	51	113	35	613	13	119
10	0	36	68	ounces.				36	630	65	95
11	0	40	20	1	17	51	113	37	648	17	71
12	0	43	109	2	35	3	89	38	665	69	47
13	0	47	61	3	52	55	65	39	683	21	23
14	0	51	13	4	70	7	41	40	700	72	136
15	0	54	102	5	87	59	17	41	718	24	112
16	0	58	54	6	105	10	130	42	735	76	88
17	0	62	6	7	122	62	106	43	753	28	64
18	0	65	95	8	140	14	82	44	770	80	40
19	0	69	47	9	157	66	58	45	788	32	16
20	0	72	136	10	175	18	34	46	805	83	129
21	0	76	88	11	192	70	10	47	823	35	105
22	0	80	40	12	210	21	123	48	840	87	81
23	0	83	129	13	227	73	99	49	858	39	57
24	0	87	81	14	245	25	75	50	875	91	33
dwt.				15	262	77	51	51	893	43	9
1	0	87	81	16	280	29	27	52	910	94	122
2	1	75	25	17	297	81	3	53	928	46	98
3	2	62	106	18	315	32	116	54	945	98	74
4	3	50	50	19	332	84	92	55	963	50	50
5	4	37	131	20	350	36	68	56	981	2	26
6	5	25	75	21	367	88	44	57	998	54	2
7	6	13	19	22	385	40	20	58	1016	5	115
8	7	0	100	23	402	91	131	59	1033	57	91
9	7	88	44	24	420	43	109	60	1051	9	67
10	8	75	125	25	437	95	85	61	1068	61	43
11	9	63	69	26	455	47	61	62	1086	13	19

TABLES OF GOLD COIN.

A TABLE

For receiving and paying the Gold Coins of Great-Britain and Portugal, at 100 cents for 27 grains, according to Act of Congress.

grs.	dol. cts.	27ths of a ct.	dwt.	dol. cts.	9ths of a ct.	oz.	dol. cts.	9ths of a ct.
1	0 3	19	12	10 66	6	28	497 77	7
2	0 7	11	13	11 55	5	29	515 55	5
3	0 11	3	14	12 44	4	30	533 33	3
4	0 14	22	15	13 33	3	31	551 11	1
5	0 18	14	16	14 22	2	32	568 88	8
6	0 22	6	17	15 11	1	33	586 66	6
7	0 25	25	18	16 00	0	34	604 44	4
8	0 29	17	19	16 88	8	35	622 22	2
9	0 33	9	20	17 77	7	36	640 00	0
10	0 37	1	ounces.			37	657 77	7
11	0 40	20	1	17 77	7	38	675 55	5
12	0 44	12	2	35 55	5	39	693 33	3
13	0 48	4	3	53 33	3	40	711 11	1
14	0 51	23	4	71 11	1	41	728 88	8
15	0 55	15	5	88 88	8	42	746 66	6
16	0 59	7	6	106 66	6	43	764 44	4
17	0 62	26	7	124 44	4	44	782 22	2
18	0 66	18	8	142 22	2	45	800 00	0
19	0 70	10	9	160 00	0	46	817 77	7
20	0 74	2	10	177 77	7	47	835 55	5
21	0 77	21	11	195 55	5	48	853 33	3
22	0 81	13	12	213 33	3	49	871 11	1
23	0 85	5	13	231 11	1	50	888 88	8
24	0 88	24	14	248 88	8	51	906 66	6
dwt.	dol. cts.	9ths of a ct.	15	266 66	6	52	924 44	4
1	0 88	8	16	284 44	4	53	942 22	2
2	1 77	7	17	302 22	2	54	960 00	0
3	2 66	6	18	320 00	0	55	977 77	7
4	3 55	5	19	337 77	7	56	995 55	5
5	4 44	4	20	355 55	5	57	1013 33	3
6	5 33	3	21	373 33	3	58	1031 11	1
7	6 22	2	22	391 11	1	59	1048 88	8
8	7 11	1	23	408 88	8	60	1066 66	6
9	8 00	0	24	426 66	6	61	1084 44	4
10	8 88	8	25	444 44	4	62	1102 22	2
11	9 77	7	26	462 22	2	63	1120 00	0
			27	480 00	0	64	1137 77	7

MERCANTILE PRECEDENTS.

BILL OF EXCHANGE.

Newburyport, Feb. 12, 1804.

EXCHANGE for £.1000 sterling.

At twenty days sight of this my first of exchange (second and third of the same tenor and date not paid) pay to John Parker, or order, One Thousand Pounds Sterling, with or without further advice from

Your humble servant,
WILLIAM PRINCE.

Messrs. Dutton & Green,
   Merchants,
      London.

BILL OF GOODS,

At an advance on the sterling cost.

Boston, May 5, 1804.

Mr. WILLIAM POOLE,

Bought of SIMON SIMMONDS,

32 ells mode	1s. 8d. sterl.	£.2 13	4
64 yds. striped Nankins	1s. 6d.	4 16	0
28 .. striped calico	1s. 9d.	2 9	0
4 pieces russel	24s.	4 16	0

Sterl. 14 14 4
Exchange 33⅓ per cent. 4 18 1¼

£.19 12 5¼
Advance at 20 per cent. 3 18 5¾

£.23 10 11

Dollars 78,48

Received his note at 2 months,

SIMON SIMMONDS.

L 2

MERCANTILE PRECEDENTS.

PROMISSORY NOTE.

Boston, May 5, 1804. For value received, I promise to pay to Simon Simmonds, or order, seventy-eight dollars forty-eight cents, on demand, with interest after two months.

Attest,   WILLIAM POOLE.
SAUL JAMES.

A RECEIPT FOR AN ENDORSEMENT ON A NOTE.

Boston, July 12, 1804. Received from Mr. William Poole, (by the hands of Mr. Benjamin Flint,) Thirty-eight dollars seventy cents, which is endorsed on his note of May 5, 1804.
                                         SIMON SIMMONDS.
38 dols. 70 cts.

RECEIPT FOR MONEY RECEIVED ON ACCOUNT.

Boston, January 10, 1804. Received from Mr. D. Evans, (by the hands of Mr. Thomas Dunmore,) Four hundred and thirty dollars on account.
   430 dols.   GEORGE PACE.

PROMISSORY NOTE BY TWO PERSONS.

Newburyport, 12th July, 1804. For value received we jointly and severally promise to pay to Mr. Samuel Rich, or order, Five hundred dollars fifty-four cents, on demand, with interest.

Attest,   NATHAN SAYBORN.
WILLIAM BOLTON.   STEPHEN NEEDY.

GENERAL RECEIPT.

New-Bedford, March 27, 1804. Received from Mr. N. B. the sum of ten dollars twenty-nine cents in full of all demands.
   10 dols. 29 cts.   E. D.

MERCANTILE PRECEDENTS.

BILL OF PARCELS.

Newburyport, June 20, 1804.

Mr. WILLIAM HOLMAN

Bought of DANIEL GREEN,

8 hhds. sugar, wt. viz.

	C. q. lb.		C. q. lb.
No. 1.	5 2 7	5.	5 3 19
2.	5 1 22	6.	5 1 17
3.	6 0 13	7.	5 1 7
4.	5 2 13	8.	5 3 14
	22 2 27		22 2 1
	22 2 1		
	45 1 0		

Tare 12 per cwt. 4 3 11

               dols. cts.

Neat 40 1 17 at 12 dols. per cwt. 484 82

2 bbls. sugar, viz.

  C. q. lb.
  2 2 25
  1 3 17

  4 2 14

Tare 21 lb. pr. bbl. 1 14

Neat 4 1 0 at 10 dols. 42 50

3 hhds. molasses, viz.

 gals.
 101—9*
 108—5
 107—7

 316—21
 21

  295 gallons at 50 cents 147 50
1 quarter cask Malaga wine 25 00
5 cases gin, at 4 dols. 25 cts. 21 25

         Dols. 721 07

* The ullage is thus noted.

MERCANTILE PRECEDENTS.

INVOICES.

INVOICE of 20 hhds. clayed sugar and 10 hhds. coffee, shipped by ...... of Boston, in the United States of America, on his own account and risque, on board the ship ...., A. B. master, bound for ...... and a market, consigned to the said A. B. for sales and returns, *viz.*

20 hhds. clayed sugar, viz.

B. C.		*C. q. lb.*		*C. q. lb.*
No. 1 a 20	No. 1.	11 3 14	11.	12 0 14
	2.	10 3 21	12.	10 2 14
	3.	11 0 0	13.	10 2 21
	4.	12 1 0	14.	11 3 21
	5.	11 1 14	15.	10 1 14
	6.	10 3 7	16.	10 2 0
	7.	10 2 0	17.	11 2 21
	8.	11 0 7	18.	10 1 14
	9.	11 0 21	19.	11 1 7
	10.	10 0 7	20.	10 3 14
		111 0 7		110 2 0
		110 2 0		
		221 2 7		
Tare 12 per cwt.		23 2 27		

                                                                         dols. cts.
197 3 8 neat, at 10 dols. 25 cts. 2027 67

10 hhds. coffee, wt. viz.

B. C.	No.	*C. q. lb.*	*Tare.*	No.	*C. q. lb.*	*Tare.*
No. 1 a 10	1.	9 2 7	108	6.	6 1 14	79
	2.	9 3 0	113	7.	6 1 6	61
	3.	10 1 21	106	8.	8 2 4	84
	4.	10 2 14	103	9.	9 1 8	91
	5.	8 0 14	94	10.	10 0 14	108
		48 2 0	523		40 2 18	423
		40 2 18	423			
		89 0 18 = 9986 lb.	———946.			
	deduct tare		946			

                          9040 lb. neat at 21 cts. 1898 40

                                                                 3926 07

Premium of insuring 4176 dols. 67 cts. at 6 per cent. } 250 60
      to cover the amount .... }

                                                         Do's. . 4176 67

Boston, &c.

INVOICE.

INVOICE of merchandize on board the brig Swan, A. B. master, shipped by A. M. on his own account and risque, for the West-Indies, and consigned to said master for sales and returns, viz.

	140 M. of boards and plank, dol.	10 dols.	1400
	20 M. of white-oak hhd. staves	30	600
	12 M. of red-oak hhd. do.	12	144
	130 M. shingles	3	390
B. No. 1--18.	18 hhds. of cod-fish, 17303 lb.	4 pr. C.	692 12
B. No. 1--52.	52 bbls. of beef	12	624
E. No. 1--30.	30 bbls. of salmon	10	300
F. No. 1----2.	2 bbls. pork	18	36
L. No. 1----7.	7 casks of rice, neat, 39 C.		
	3 qrs. 21 lb............	4pr.cwt.	159 75
	3 M. of hoops	25	75
	1300 pair of shoes	50 cts.	650

Dols. 5070 87

Portsmouth, Sept. 7, 1804.
Errors excepted.

A. M.

Mr. Abraham Jones, to Walter Brown   Dr.

1804.
Jan.	5.	For 1 barrel of flour		Dols. 10	
	8.	4 lb. coffee .... 2s.		....	1 33
	9.	9 lb. of sugar...........	11d.	....	1 37
	23.	7 gallons of molasses .... 3s.	9d.	....	4 37
Feb.	7.	3 quintals of fish .... 15s.		....	7 50
	16.	2 lb. hyson tea .... 8s.	6d.	....	2 83
Mar.	29.	5 lb. chocolate .... 1s.	6d.	....	1 25
May	5.	2 bushels of corn .... 4s.	9d.	....	1 58

Dols. 30 23

Errors excepted.

ACCOUNTS OF SALES.

SALES of 20 hhds. 7 bbls. and 31 bags coffee, for and on risk of Mr. William Stillman, merchant in Portland.

1804.
March 15	William Edes, 20 hhds. wt. 14,376 lb. at 23 cts. per lb.		Dols.	3306	48
16	George Watts, 7 bbls. wt. 1493 at 23 cts.			343	39
17	Peter Bates, 31 bags, 5507	23		1206	61

Charges,			4916 48
Advertising	Dol. 1 46		
Storage	3 50		
Commission on 4916 dols. 48 cts. at 2½ per cent.	122 91		127 87

Neat proceeds passed to his credit Dols. 4788 61
Errors excepted, &c.

SALES of sundry merchandize received per the ship Juno, Capt. Dane from Machias, and disposed of for account and risk of Amos Goodwin, merchant there.

Date	To whom sold	quintals fish	barrels oil	barrels salmon	barrels herring	cords wood	cords bark	feet boards	barrels beef	Price	Amount
										dls.cts.	dols.cts.
1804.											
June 4	James Yates	30								3	90
8	Wm. Rowe	120								3 27	292 40
27	John Payson		6							12	72
July 4	James Nugent				22					4	88
··	Cash			50						8 75	437 50
8	Sim. Sands							3,216		6 50	20 90
21	Stock								15	9	135
29	Paul Simson				18					3 50	45 50
Aug. 5	Jona. Rowe							1,259		6	7 55
	Taken to fill up		1								
		150	7	50	18	22		4,475 15			1288 85

Remaining unsold, 40 barrels of herring.
Charges, viz.
Storage of fish · Dols. 10 50
Commisson on 1288 dols. 85 cts. at 2½ per cent. 32 22   42 72

Neat proceeds carried to the credit of his account, Dols. 1246 13
Errors excepted, &c.

SALES *of* 19 *hogsheads and* 7 *barrels of rum, received per the schooner Ruby, Richard Butler, master, from Portsmouth, for account and risk of Daniel Edwards, merchant there.*

Date.	To whom sold.	19 hhds. Rum.	7 bbls. Rum.	Gallons.	Price.	Contents.	Amount.
1804.					Cts.		dols. cts.
May 24	By Walter King		1	29½	100		29 50
June 2	By David Jones	2		216	100	110 and 106	216
20	By James Ray	4		438	96	108,110,111,109	420 48
24	By Aaron Judson		3	81	95	26¼, 27¼, 27	76 95
July 23	By Tho's Ropes	1		115	95½		109 82
Aug. 3	By Parsons & Ely		1	25	95½		23 87
23	By Simon Sands	2		222	98	109, 113	217 56
Sept. 4	By Miles Young	1	1	138	96	110, 28	132 48
10	By Moses Bliss	3	1	342½	99	107,104,103,28½	339 7
25	By Amos Dundas	6		632	98¼	109,102,106 111, 112, 92	622 52
		19	7	2239			2188 25

Charges.

	dls. cts.	dls. cts.
Paid Capt. Butler freight of 19 hhds. rum, at	2 50	47 50
ditto ... 7 bbls.	66	4 62
Porterage 19 hhds.	40	7 60
ditto 7 bbls.	10	70
Gauging 26 casks	12½	3 25
Cooperage 3 dols. on hhds. 1 dol. 50 cts on bbls.		4 50
Advertising		1 25
Commission on 2188 dols. 25 cts. at 5 per cent.		109 41
		178 83

Neat proceeds··Dols. 2009 42

Outstanding in hands of

                     dls. cts.
Moses Bliss ·········339 7
Amos Dundas ·······622 52

Boston, 25th *September,* 1804.

Errors excepted, &c.

MERCANTILE PRECEDENTS.

SALES of the ship Hiram's Cargo, by William Sutton.

1804.	lb.	liv.	sol. den.	liv. sol. den.
May 24. 65 hhd. fish, wt. nt. 72587 at 33 liv. per 100,		23953	14 2	
6 do. do. 6515 32		2084	16 0	
2 do. do. 2136 31		662	3 2	
34 do. do. 36658 30		10997	8 0	
2 do. partly dam. 2184 sold at auction for 226			0 0	
				37924 1 4
109				

liv. sol. den.

24 bbls. beef, at 101 1 3 per bbl.	2425 10 0	
7 do. do. 99 8 5	695 18 11	
29 do. do. 90 15 0	2631 15 0	
4 do. do. 83 0 0	332 0 0	
		6085 3 11
64		

liv. sol.

13 bbls. pork 136 0	1768	0 0	
25 do. porter 80 0	2000	0 0	
3 box. lin. con. 169 piec. 96 0 pr. pie.	16224	0 0	
14 firk. butter, wt. 1129 lb. 2 5 pr. lb.	2540	5 0	
5 thousand hoops .... 240 pr. M.	1200	0 0	
59 do. shingles .... 16 do.	944	0 0	
15949 feet boards .... 120 do.	1913	17 7	
170 shaken hhds. .... 8¼ pr. hhd.	1402	10 0	
			27992 12 7

	liv. s. d.		72001 17 10
Commission on 72001 17 10 at 5 per cent.			3600 1 10
		Liv.	68401 16 0

Errors excepted, &c.

Disbursements, Duties, &c. paid on ship Hiram, by Wm. Sutton.

1804.	liv. s. d.	liv. s. d.
May 18. Paid for a barrel of flour	86 10 0	
.... to the admiralty	240 11 6	
... for fresh meat	56 12 5	
.... for flats to unload with	341 13 6	
		725 7 5
Paid to the harbour master	66 10 4	
.... for storage and negro hire	619 14 8	
.... for inward duties...............	714 11 7	
.... for outward duties	229 13 5	
		1630 10 0
Paid for brokerage	821 13 6	
.... for passport and certificate	68 19 7	
		890 13 1

Point-Petre, Guadaloupe, July 12, 1804.

Liv. 3246 10 6

Errors excepted, &c.

Wm. SUTTON.

MERCANTILE PRECEDENTS.

Dr.   Mr. *William Cummins*, as owner of the Ship *Hiram*, in account with *William Sutton*.   **Cr.**

1804.	hhd.	lb.	liv.	lb.	liv. s. d.	liv. s. d.	1804.		liv. s. d.	liv. s. d.
June 12.	To 24 sugar, wt. nt. 35343 at 58 pr. 100,				20498 18 9		June 10.	By neat proceeds of ship Hiram's cargo, per account of sales annexed.........		68401 16 0
	7 do. .... 9055..42				3803 2 0			By cash brought out		1750 0 0
	12 do. .... 19983..43				8592 13 9					
	—									
	43	64381				32894 14 6				
		sol. d.								
	To 4856 lb. coffee, at 22 6 per lb.			5463 0 0						
	2019 ....do..... 23			2321 17 0						
	6523 ....do..... 23 6			7664 10 6						
	5247 ....do..... 24			6296 8 0						
	18645 in 20 hhds. and 6 bbls.					21745 15 6				
To 19 bales cotton 4645 lb. at 140 liv. per 100,				6503 0 0						
To 2661 velts of molasses, at 24 sols per velt,				3193 4 0						
						9696 4 0				
To commission on 64336 liv. 6 s. 4 d. at 2½ per cent.						1608 8 4				
To amount of disbursements, duties, &c. per account annexed						3246 10 6				
						69191 12 10				
		Balance in cash on board				960 3 2				
				Livres..70151 16 0						

Livres..70151 16 0

Point-Petre, Guadaloupe, July 12, 1804.
Errors excepted.
WILLIAM SUTTON.

Z

Dr. Mr. John Johnson in account current with William Roberts. Cr.

1803.		dols. cts.	1803.		dols. cts.
May 19.	To cash advanced per receipt	300	Oct. 28.	By ship Columbia for hull thereof complete, being 171½ tons, at 16 dls.	2744
June 5.	To sui lmes per bill	458 12			
July 25.	To payment of his order to M. B. for	100			
29.	To 1 bag coffee 96 lb. at 20 cents	19 20			
Ag. 1.	To cash per receipt	430			
—	To 5 hhds. rum 555 gals. at 83⅓ cents	462 50			
—	To 3 boxes glass 7 by 9, 10 dols.	30			
Sept. 2.	To sundries per bill	228 56			
—	To cash per receipt	385			
20.	To 12 bl. flour at 8 dols. 4 bl. pork at 12	144			
25.	To 1 hhd. sugar 8 cwt. 2 qrs. 7 lb. nt.	85 62			
—	10 dols.				
Oct. 29.	To cash and sundries in full	81			
		dols. 2744			dols. 2744

Salem, October 28, 1803.
Errors Excepted.
WILLIAM ROBERTS.

MERCANTILE PRECEDENTS.

Mess. Wilson & Gale in account current with William Duncan.

Dr. 1804.		dols. cts.
May 19.	To 1 barrel flour .... delivered Wilson	9 50
28.	To cash .... Gale	28 50
June 22.	To 15 lb. butter 15s. tea 6s. .... Wilson	3 50
29.	To 3 quintals scale fish, at 15s. Gale	7 50
July 2.	To 1 half barrel flour .... do.	5
15.	To 12 lb. coffee 18s. 28 lb. sugar 25s. 6d. do.	7 25
20.	To cash .... Wilson	25
26.	To 15 bushels corn at 5s. ...... Gale	12 50
Aug. 2.	To 8 do. rye at 6s. ...... Wilson	8
	To paid their order to James Rowe	12 50
		dols. 119 25

Cr. 1804.		dols. cts.
Sept. 10.	By amount of their account for repairs on the ship America	119 25
		dols. 119 25

Newburyport, 27th. Sept. 1804.
Errors excepted.
For Mr. WILLIAM DUNCAN.
SAMUEL TRUSTY.

MERCANTILE PRECEDENTS.

Dr.   Mr. James Richardson, in account current with Thomas Seccome.   Cr.

1803.		dols. cts.	1803.		dols. cts.
June 12.	To sundries per bill................	28 26	Oct. 12.	By schooner William for blacksmith's work per bill, viz. 6325lb. at 6c. 759lb. tare at 12 per ct. 7084 lb. supplied per Dr.	379 50
	To 53 bars Iron, wt. 21 2 10 for schooner William				
July 15.	To 121 do........41 2 18 do. do.				
26.	To 1 hhd. W.I. Rum qt. 107 gals, at 96 cts.	102 72			
Aug. 28.	To Cash per receipt...............	180 00			
29.	To 4 bbls. flour at 9 dols. 50 cts.	38 00			
Sept. 21.	To cash paid his order to James Wise.....	28 00			
	To cash in full	2 52			
		dols. 379 50			dols. 379 50

Newburyport, 12th October, 1803.
Errors excepted.
                      THOMAS SECCOME.

NOTE. *When a person is furnished with his account current, it is necessary to specify the various charges, and when they are numerous, some accountants make but one charge of them, in the account current, referring to an annexed account of the several articles thus included.*

BILL OF SALE.

TO all people to whom this present Bill of Sale shall come, I R. P. of Newburyport, in the State of Massachusetts, Merchant, send Greeting; KNOW YE, That I the said R. P. for and in consideration of the sum of three thousand, two hundred and twenty-two dollars, to me in hand well and truly paid at or before the ensealing and delivery of these presents, by S. T. of the said Newburyport, Merchant, the receipt whereof I do hereby acknowledge and am therewith fully and entirely satisfied and contented, have granted, bargained and sold, and by these presents do grant, bargain and sell, unto the said S. T. all the hull or body of the good brig Sally, together with all and singular her masts, spars, sails, rigging, cables, anchors, boats and appurtenances, now lying at Newburyport, and registered at the port of Newburyport, the certificate of whose registry is as follows:

IN pursuance of an Act of the Congress of the United States of America, entitled, " An ACT concerning the registering and recording of ships or vessels," R. P. of Newburyport, in the State of Massachusetts, Merchant, having taken or subscribed the oath required by the said act, and having sworn that he is the only owner of the ship or vessel called the Sally, of Newburyport, whereof William Knapp is at present master, and is a citizen of the United States, as he hath sworn, and that the said ship or vessel was built at Salisbury, in the said state, in the year seventeen hundred and ninety-nine, as also appears by a certificate of enrolment, No. 129, issued in this district on the fourth day of August last, now surrendered,—and N. S surveyor of this district, having certified that the said ship or vessel has one deck and two masts, and that her length is sixty-nine feet five inches, her breadth twenty-two feet and one half inch, her depth eight feet two inches, and that she measures one hundred and six tons and forty ninety-fifths, that she is a square sterned brig, has no galleries and no figure head, and the said R. P. having agreed to the description and admeasurement above specified, and sufficient security having been given according to the said act, the said brig has been duly registered at the port of Newburyport.

Given under my hand and seal at the port of Newburyport, this first day of January, in the year one thousand eight hundred.

To have and to hold the said granted and bargained brig Sally and premises with the appurtenances, unto the said S T. his heirs, executors, administrators or assigns to his only proper use, benefit and behoof forever. And I the said R. P. do avouch myself to be the true and lawful owner of the said brig and appurtenances, and have in myself full power, good right and lawful authority to dispose of the said brig as aforesaid, and her appurtenances in manner as aforesaid, and furthermore I the said R. P. do hereby covenant and agree to warrant and defend the said brig and premises, with the appurtenances against the lawful claims and demands of all persons whatsoever unto the said S T. In witness whereof, I the said R. P. have hereunto set my hand and seal, this tenth day of June, in the year of our Lord one thousand eight hundred.

Dr.　　　　　*Mr. Thomas Gibson in interest*

	dol. cts.		days.	dol. ct.
To Int. on	35 00	fr. Jan. 31, '96 to Oct. 12,'96,	256	1 47
To do. on	2962 19	..Feb. 2....to..do.......	254	123 68
To do. on	2590 42	..May 31,....to..do.......	134	57 06
To do. on	1733 97	..July 2....to..do.......	102	29 07
To do. on	73 63	..July 12....to..do.......	92	1 11
To do. on	455 52	..Aug. 25....to..do.......	47	3 51
To do. on	158 71	..Sep. 30....to..do.......	12	0 31

　　　　　　　　　　　　　　　　　dols.　216 21

Dr.　　　　　*Mr. William Mace in interest*

1798.		dols. cts.	y. m. d.	dols. cts.
March 3. To Interest on	3869 20	for 1	5 11	335 97
April 26............on	273 6	..1	3 18	21 29
Aug. 18............on	400	..	11 26	23 73
Dec. 28............on	414 6	..	7 16	15 59
'99 Ja.15............on	200	..	7 9	7 30
Feb. 19............on	300	..	5 25	8 75
Mar. 26............on	1300	..	4 18	29 90

　　　　　　　　　　　　　　　　　dols.　442 58

Account with Thomas Merchant   Cr.

	dols. cts.		days.	dols.ct.
By interest on 500	from Apr. 24,'96, to Oct. 12,'96,		171	14 5
By do.	1133 25 ...... 25 12,		170	31 67
By do.	296 24 May 3 12,		162	7 88
By do.	215 ...... 5 12,		160	5 65
By do.	215 80 June 9 12,		125	4 43
By do.	109 74 ...... 24 12,		110	2 0
By do.	517 90 July 20 12,		84	7 15
Balance due on this account carried to the debit of ac't.				143 38
			dols.	216 21

Salem, &c.

---

account with Thomas Merchant   Cr.

1799.		dols. cts.		dols. cts.
Jan. 16.	By interest on	339 67		
		427 81		
		———	Y. m. d.	
		767 48	— 6 18	25 32
Balance carried to account current				417 21
			dols.	442 53

Salem, August 26th, 1799.
         Errors excepted,
                 THOMAS MERCHANT.

CHARTER-PARTY.

THIS Charter-party of affreightment, indented, made and fully concluded upon this ninth day of June, in the year of our Lord, one thousand eight hundred, between J P. of Boston, in the county of Suffolk, and Commonwealth of Massachusetts, merchant, owner of the good ship Helen, of the burden of two hundred tons, or thereabouts, now lying in the harbour of Boston, whereof R. P. is at present master, on the one part, and C. D. of said Boston, merchant, on the other part, *Witnesseth*, That the said J. P. for the consideration hereafter mentioned, hath letten to freight the aforesaid ship, with the appurtenances to her belonging, for a voyage to be made by the said ship to London, where she is to be discharged (the danger of the seas excepted) and the said J. P. doth by these presents covenant and agree with the said C. D. in manner following, *That is to say*, That the said ship in and during the voyage aforesaid, shall be tight, staunch and strong, and sufficiently tackled and apparelled with all things necessary for such a vessel and voyage; and that it shall and may be lawful for the said C. D. his agents or factors, as well at London as at Boston, to load and put on board the said ship, loading of such goods and merchandize as they shall think proper, contraband goods excepted.

IN consideration whereof, the said C. D. doth by these presents, agree with the said J. P. well and truly to pay, or cause to be paid, unto him, in full for the freight or hire of said ship and appurtenances, the sum of three dollars per ton, per calendar month, and so in proportion for a less time, as the said ship shall be continued in the aforesaid service, in sixty days after her return to Boston. And the said C. D. doth agree to pay the charge of victualing and manning said ship and all port charges and pilotage during said voyage, and to deliver the said ship on her return to Boston, to the owner aforesaid or his order. And to the true and faithful of all and singular the covenants, payments and agreements aforementioned, each of the parties aforenamed binds and obliges himself, his executors and administrators, in the penal sum of two thousand dollars firmly by these presents. In witness whereof, the parties aforesaid have hereunto interchangeably set their hands and seals the day and year afore-written.

BILL OF LADING.

J. R.
1 a 53
Casks Potash.
ton cwt.
8 18   *l. s. d.*
at 80s.—35 12 0
Primage 5
pr. ct.   1 15 7
―――――
£.37 7 7
―――――

SHIPPED in good order and well conditioned by John Rolly, in and upon the good ship called the Iris, whereof is master for this present voyage Charles Ely, and now riding at anchor in the harbour of Newport, and bound for Liverpool, to say, *fifty three casks of potash, containing eight tons and eighteen cwt* being marked and numbered as in the margin, and are to be delivered in the like good order and well conditioned, at the aforesaid port of Liverpool (the danger of the seas excepted) unto Mr. J. May or to his assigns, they paying freight for the said goods, *four pounds* British sterling per ton, with five per cent. primage. In witness whereof, the master or purser of the said ship hath affirmed to three bills of lading all of this tenor and date, the one of which being accomplished, the other two to stand void. Dated in Newport, July 7th, 1804.   C. ELY.

EXCHANGE.

PALERMO IN SICILY.

Accounts are kept in Onges, Tarins and Grains.

20 Grains ....make.......... 1 Tarin.
30 Tarins .... 1 Onge or Once.

Feb. 3, 1803, the value of the money of Palermo in U. S. currency was as follows:

1 Grain ....equal to............ 4 Mills.
20 do. = 1 Tarin .... 8 Cents.
240 do. = 12 do. = 1 Sc. dollar..=.. 96 do.
600 do. = 30 do. = 2½ do. = 1 Onge = 240 do.

The Spanish dollar is current at 252 grains. The value of the onge at par is 11s. 3d. sterling. The exchange on London Feb. 3, 1803, was 56 tarins for the £. sterling, or 10s. 8½d. sterling per onge.

The Cantar of Sicily = 176 lb. Avoirdupois.
The Rottoli .... = 1¾ lb. do.
100 Rottoli make a Cantar.

A Cantar of Oil is 25 gallons English measure. The Sicilian barrel contains 9 gallons.

Mahogany is sold by weight; one foot board measure will weigh about 2 rottoli.

The measure called Caffis is 3¼ gallons.
The lb. in Sicily is 12 oz. avoirdupois.
The Salm is 485 lb. avoirdupois.

Examples.

1. What cost 264 Cantars 25 rottoli of Mahogany at 8 onges 15 tarins per cantar?

```
              264
                8
             ----
             2112
15 tar. ½ =  132
25 rot. ¼ =    2  3 15
             ---------
             2246 3 15
```

Ans. 2246 ong. 3 tar. 15 gr.

EXCHANGE.

2. A cargo consisting of 3564 quintals of Fish invoiced at 5 dols. 50 cts. per quintal, is sold in Palermo at 75 per cent. advance; what sum must be received for it at 252 grains per dollar?

```
                          3564
                             5
                         ─────
                        17820
        50 cts. ½ =     1782
                        ──────
                        19602
        50 per ct. ½ =   9801
        25 ...... ¼ =   4900 50
                        ──────
                dols.  34303 50
                         252
                        ──────
                        68606
                       171515
                        68606
        50 cts. ½ =      126
                       ────────
               2|0)86444 8|2 grains.
                      ─────────
               3|0)43222|4  2
                      ─────────
                      14407 14 2
```
Ans. 14407 ong. 14 tar. 2 gr.

3. What is the Brokerage on 13131 ong. 12 tar. at 1⅛ per cent?

```
                      13131 12
                             1
                      ────────
                      13131 12
         ⅛  =          1641 12 15
                      ──────────
                      147|72 24 15
                           30
                      ──────────
                           21|84
                              20
                      ──────────
                           16|95
```
Ans. 147 ong. 21 tar. 16 gr.

Newburyport Navigation Book, Chart, and Stationary Store.

EBENEZER STEDMAN,

HAS CONSTANTLY FOR SALE,

Wholesale and Retail, at No. 6, State-Street, *Newburyport—*

A Compleat assortment of Navigation Books, Charts, School-Books, Stationary, and Fancy Articles, viz.

Bowditch's New American Practical Navigator ; New and improved edition of the American Coasting Pilot ; Marshal on Insurance ; Abbot on Shipping ; Morse's Universal Geography ; Do. abridged ; Morse's American Gazetteer ; Morse & Parish's Gazetteer of the Eastern Continent ; Morse's Gazetteer, abridged , Murray's Grammar; Do. abridged ; Murray's English Reader ; Do. Exercises ; Beauties of the Bible ; Columbian Orator ; Webster's 1st and 3d Parts ; Perry's Spelling Book ; Young Man's Best Companion ; Johnson's, Walker's, Barclay's, Bailey's, and Perry's English Dictionaries ; Johnson's do. in miniature; Ainsworth's and Young's Latin and English Dictionaries ; Boyer's French do. ; American Preceptor : Dwight's Geography ; Art of Reading ; Morse's Elements of Geography ; Fiske's Spelling Book ; Child's Companion; Ladies' Accidence ; Primers; Youth's Library ; and School and Classical Books of every kind.

Handsome Folio Bibles, with plates ; royal quarto do. ; Oxford and Edinburgh do. with and without Apocrypha ; Octavo do. ; elegant and common pocket do. ; Oxford and Edinburgh school do.: Testaments, large print ; Church Prayer Books, elegant and common ; Belknap's Psalms and Hymns, morocco and common bindings ; Watts' do. large print, do. in miniature ; Smith and Sleeper's Hymns, &c. &c.

STATIONARY of every description, *viz.* Medium, demy, thick and thin folio post, foolscap and quarto post English Paper, of every kind ; foolscap and pot American do. of various qualities ; Bonnet Paper ; Wrapping Paper ; Quills, of every quality and price ; Slates, of all sizes ; Wafers ; Sealing Wax ; red and black Ink Powder ; lead, slate, and hair Pencils ; Copy Books ; Inkstands, of all kinds ; boxes of Paints ; Penknives ; India Rubber ; Playing Cards ; Ledgers ; Journals ; Waste and Record Books, of every kind ; Cyphering and Writing Books ; Memorandum Books ; Charts, Pilots, Quadrants, Spy Glasses, Log Books, &c. &c.

✶✶✶ *A large variety of BOOKS on all subjects kept for Circulation at the above Store, where constant attendance will be given, and every favour gratefully acknowledged.*

☞ Cash given at the above store for RAGS.

Sept. 1802.

BOOK-STORE,

At the New Fire proof Brick Store, near the head of *Fish-street*, PORTLAND, and nearly opposite the *Insurance Offices.*

THOMAS CLARK,

BOOKSELLER & STATIONER,

KEEPS constantly for sale, at the *Boston prices*, a general assortment of Bibles, all sizes; Testaments—Dictionaries—Webster's 1st, 2d & 3d Parts—Maine Spelling Book—American Preceptor—Ladies' Accidence—Child's Companion—Morse's Universal Geography—Nichols', Pike's and Walsh's Arithmetics—Art of Reading, &c. &c.

ALSO,

A general assortment of Account Books, Receipt and Memorandum Books—Music Books, of all kinds—Playing Cards—Black and Red Inkpowder—Writing and Cyphering Books.—*Likewise.*

A compleat assortment of Seamen's Books and articles; among which are, Blunt's New Practical Navigator—American Coast Pilot, new edition—Ship Master's Assistant, &c. &c.—Hadley's Quadrants, different prices—Gunter's Scales and Dividers—Quarter Waggoners—and an assortment of Charts.

Valuable Patent Medicines, for sale by appointment at said store, and at no other in Portland.

Book-Binding done in the neatest manner, with dispatch.

Social and Private Libraries will be supplied on better terms than at any other Store in the District of Maine.

Just published as above, wholesale & retail—Divine Hymns, or Spiritual Songs; for the use of Religious Assemblies, and private Christians: being a collection by *Joshua Smith* and *Samuel Sleeper.* To which are added Thirty two Hymns.

☞ Lottery Tickets, generally for sale at the above store.

HARVARD COLLEGE LIBRARY

THE ESSEX INSTITUTE
TEXT-BOOK COLLECTION

GIFT OF

GEORGE ARTHUR PLIMPTON

OF NEW YORK

JANUARY 25, 1924

Lightning Source UK Ltd.
Milton Keynes UK
UKHW012205201118
332686UK00007B/115/P